Arts and Humanities Dat

Creating and Using *Virtual Reality*

Arts and Humanities Data Service

Creating and Using *Virtual Reality*: *A Guide for the Arts and Humanities*

Kate Fernie and Julian D. Richards

with contributions by
Tony Austin, Rachael Beach, Aaron Bergstrom, Sally Exon,
Marc Fabri, Michael Gerhard, Catherine Grout, Stuart Jeffrey,
Mike Pringle, Damian Robinson, Nick Ryan, Melissa Terras and
additional Case Studies by Kate Allen, Clive Fencott,
Learning Sites and Anthony McCall

Oxbow Books 2003

Published by Oxbow Books for the Arts and Humanities Data Service

ISBN 1 84217 040 6
ISSN 1463 - 5194

A CIP record of this book is available from the British Library

This book is available direct from

Oxbow Books, Park End Place, Oxford OX1 1HN
(Phone: 01865-241249; Fax: 01865-794449)

and

The David Brown Book Company
PO Box 511, Oakville, CT 06779, USA
(Phone: 860-945-9329; Fax: 860-945-9468)

or from the website
www.oxbowbooks.com

Printed in Great Britain by
Information Press, Eynsham, Oxford

Contents

Acknowledgements

The authors and editors of this *Guide* would like to offer sincere thanks to those who have contributed to it by reviewing and commenting on draft versions. These people include:

- Jo Clarke, Archaeology Data Service
- David Dawson, Re:source
- Alligator Descartes, Symbolstone
- Mark Gillings, University of Leicester
- Dave Hobbs, Bradford University
- Jeremy Huggett, Glasgow University
- Hamish James, History Data Service
- William Kilbride, Archaeology Data Service
- Paul Miller, UKLON
- Keith Westcott, Archaeology Data Service
- Paul Wheatley, Camileon Project, University of Leeds

The authors and editors of this Guide would also like to extend their thanks to all those who have worked so hard to bring it to publication. These people include:

- Catherine Hardman, Archaeology Data Service
- Val Kinsler, 100% Proof
- Maureen Poulton, York St John College
- Phill Purdy, Visual Arts Data Service
- Janine Rymer, Visual Arts Data Service
- Lara Whitelaw, Visual Arts Data Service

Section 1: Overview and Objectives

1.1 WHO IS THIS GUIDE FOR?

Creating and Using Virtual Reality is intended for those who are interested in how virtual reality can be used within the arts and humanities. This *Guide to Good Practice* concentrates on accessible desk-top virtual reality which may be distributed and viewed on-line via the World Wide Web. It is concerned with the variety of virtual reality models that may be produced and how to ensure that these can be delivered successfully to users and preserved for future reuse.

This Guide introduces virtual reality by considering its history, philosophy and theory and discusses good practice in planning virtual reality projects. It does not attempt to cover all virtual reality technologies – this is a rapidly developing field and new methods are continually emerging. The techniques that are introduced are those which are being used in the Arts and Humanities and for which standards are emerging. Ensuring that the models produced can be used and enjoyed by the audiences for which they are intended is the most important consideration for virtual reality projects. The data, management and documentation procedures required to enable models to be maintained and to continue to be enjoyed are introduced in this Guide. Preservation in the longer term is an emerging field; this Guide explores strategies for archiving and considers how to avoid the loss of virtual reality models as technology changes.

A virtual library of case studies is presented illustrating some applications of virtual reality in Archaeology, Architecture, Dance, Design, Fine Art, Heritage, History, Museum Studies and Theatre. Examples of worlds which allow users to interact with each other are also presented.

A wide range of organisations and individuals are both creating and holding virtual reality models. For this reason the *Guide to Good Practice* is aimed at:

* **Creators of virtual reality** including artists, illustrators and computer scientists
* **Organisations and individuals commissioning virtual reality** including national agencies, university-based projects and funding bodies
* **Curators who will receive virtual reality models** including museums, galleries and archives.

1.2 INTRODUCING VIRTUAL REALITY

Virtual reality has been defined in many different ways. It is generally agreed that the essence of virtual reality lies with computer-based three-dimensional environments. Often termed 'worlds', they represent real-world or conceptual environments that can be navigated through, interacted with and updated in real-time. The definition of virtual reality is explored in more detail in Section 2 of this Guide.

Virtual reality can be delivered using a variety of systems. The 'world' may be projected inside a 'cave' within which users can move around. Headsets and gloves may be worn so that users are immersed in a virtual world which they can move around and touch. The most widely used form of virtual reality in use today is desk-top virtual reality. In these systems virtual reality worlds run on users' desk-top computers and are displayed on a standard monitor and navigated using a mouse or 3-D space ball with a keyboard.

Desk-top virtual reality systems can be distributed easily via the World Wide Web or on CD and users need little skill to install or use them. Generally all that is needed to allow this type of virtual reality to run on a standard computer is a single piece of software in the form of a viewer. Desk-top virtual reality is both very accessible and widely used, and for these reasons will form the focus of this Guide.

Uses of virtual reality

There are many common applications for virtual reality. They fall into the main categories of training, education, simulation, visualisation, conceptual navigation, design and entertainment but there is much overlap between these categories:

- Training applications include allowing users to practise a process repeatedly in a no-risk environment. For example, users might dig an archaeological site, trying out different strategies without the risk of destroying important evidence
- Educational applications include virtual visits and simulations. For example, a virtual visit to a museum that is too far away to visit or does not exist in the real world. Historic battles may be simulated allowing users to see 'what would have happened if?'
- Visualisation examples include an architect's design for a building or the reconstruction of ancient buildings from archaeological evidence. Such models also allow users to explore something too large or too small to explore in reality and can bring historical time-lines to life
- Applications of virtual reality for conceptual navigation enable, for example, users of a library or archive to find the information they need at a logical or physical level
- Virtual reality allows designs to be visualised and tested. For example, a design application might allow a choreographer to see a dance in action
- Entertainment applications include virtual art galleries and games. Virtual reality may also be considered as an art form in its own right
- Collaborative Virtual Environments (CVEs) allow users to interact with each other in a virtual world allowing the development of virtual communities, thus adding a new dimension to virtual reality.

1.3 HOW TO USE THIS GUIDE

Ideally, anyone involved with or planning a virtual reality project would read this Guide in its entirety. However, in many cases individuals will be involved in different stages of the project. To reflect this, the Guide has been structured into clear thematic sections and does not necessarily have to be read in a linear order. However, if you are reading about this subject for the first time, we recommend you follow the sections in order to gain most benefit.

The sections of this guide are:

Section 1: Overview and Objectives – introduces the Guide, virtual reality, the subjects and issues covered.

Section 2: Virtual Reality: History, Philosophy and Theory – charts the history of virtual reality and considers the philosophy and theory behind it.

Section 3: Virtual Reality Methods and Techniques – provides information about the creation of virtual reality. General techniques and issues are introduced and specific types of VR are presented and compared.

Section 4: Collaborative Virtual Environments (CVEs) – this section considers virtual environments where users can interact with each other. It explains the technology involved, how users may be represented and ways in which they may interact with each other.

Section 5: Documenting Data from a Virtual Reality Project – looks at the documentation needed to support testing, maintenance and archiving.

Section 6: Archiving Virtual Reality Projects – looks at archiving issues. With technology changing fast and new VR formats emerging regularly, it is important to consider ways of ensuring the use of VR resources in the future. Section 6 introduces the techniques that may be used to extend the lifespan of resources and provides guidelines for depositing virtual reality projects in digital archives.

Section 7: Resource Discovery – considers resource discovery and the reasons for providing metadata to support the retrieval of virtual reality both on the Internet and from archive catalogues.

Virtual Reality Case Study Library – this section offers examples of the uses of virtual reality in the Arts and Humanities.

This guide aims to provide guidance for those without a technical or computing background. For this reason the language used is relatively free from technical jargon and a glossary is provided to explain the more technical terms. A bibliography provides references to more in-depth information about the subjects covered in the Guide and author biographies are included for reference.

1.4 OTHER GUIDES IN THIS SERIES

Creating and Using Virtual Reality: A Guide for the Arts and Humanities has been produced by the Arts and Humanities Data Service centres for Archaeology and the Visual Arts as part of the Arts and Humanities Data Service *Guides to Good Practice* series. These guides comprise a comprehensive, authoritative and highly complementary set of practical guidelines.

Guides produced by the **AHDS Centre for Archaeology** include:

- Archiving Aerial Photography and Remote Sensing Data
- Archaeological Geophysics: a Guide to Good Practice
- CAD: Guide to Good Practice

- Digital Archives from Excavation and Fieldwork: a Guide to Good Practice
- GIS: Guide to Good Practice

The guides produced by the AHDS Centre for Archaeology in this series concentrate either upon archiving digital data produced by specific techniques, such as the products of Aerial or Remote Sensing, CAD or Geophysical surveys, or upon analytical and data management techniques, such as Geographical Information Systems.

Guides produced by the **AHDS Centre for the Visual Arts** include:

- Creating Digital Resources for the Visual Arts: Standards and Good Practice
- Using Digital Information in Teaching and Learning in the Visual Arts
- Investing in the Digitisation of Visual Arts Material (forthcoming)

The guides produced by the AHDS Centre for the Visual Arts in this series concentrate on providing an introduction to creating and developing digital resources for the visual arts, including their use to enhance the teaching and learning process.

The **Arts and Humanities Data Service** caters for digital archiving needs across the humanities disciplines of archaeology, history, performing arts, visual arts and literature, languages, and linguistics. The most up-to-date information on the other guides in the AHDS series is available on-line (http://ahds.ac.uk/guides.htm) but titles include:

Title	Centre
Creating and documenting electronic texts	Literature, Languages, and Linguistics
Developing linguistic corpora	
Finding and using electronic texts	
Digitising history: a guide to creating digital resources from historical documents	History
A Place in History: A Guide to using GIS in Historical Research	
Creating digital performance resources	Performing Arts
Creating digitised audio materials for use in research and teaching	
Digital Collections in the Performing Arts: Metadata, Management and Minefields	

Table 1: Other Guides to Good Practice *from the Arts and Humanities Data Service*

These guides provide information about how to prepare and deposit digital material in a digital archive. Although they give some recommendations concerned with the facilities and procedures for the creation and maintenance of a digital archive, these are covered more fully in the Arts and Humanities Data Service's *Managing Digital Collections* publications.

Section 2: Virtual Reality: History, Philosophy and Theory

This chapter presents a background to the technologies which have become known as Virtual Reality (VR). It begins with a discussion of what VR actually is, taking a brief look at the term itself and the origins of the technology, then gives an outline of VR's main components and attributes and describes its breadth of use. The chapter then explores theoretical and philosophical aspects that may be taken into consideration by developers of VR applications, in particular the constant necessity to bear in mind the needs, wants and perspectives of the intended user.

2.1 WHAT IS VIRTUAL REALITY?

Broadly, virtual reality (VR) is the label given to a range of computer-based approaches to the visualisation of concepts, objects or spaces in three or more dimensions. Although the distinction is becoming increasingly blurred, these approaches tend to differ from other three-dimensional visualisations, such as the output by Computer Aided Design (CAD) packages and Geographic Information Systems, in that the experience is interactive. The user of VR is often able to move around within the three-dimensional space and may be able to interact with objects found there.

There are many different ways of interpreting the term *virtual reality*. It can be seen as a technology that enables interaction with 'three-dimensional databases' or as a way of 'integrating man with information' (Stone 1998; Warwick *et al.* 1993). This idea of information being at the core of VR is supported by those who promote VR as a method of transferring knowledge or of turning information into knowledge, for example about a route, area or other virtual space (Witmer *et al.* 1996; Machover and Tice 1994). The most straightforward description of VR is perhaps the military notion of synthetic environments.

The definition of the term VR itself is unimportant, except when it promotes unhelpful expectations in the mind of a potential user. For example, the media-hyped image of VR is as surreal, artificial worlds into which participants are immersed via various futuristic gadgets. Putting this image aside, VR is rapidly developing into a practical and powerful imaging tool for a wide variety of applications. Consequently, when developing VR projects it may be wise to avoid definitions and concentrate on what it is that the technology can do.

2.2 ORIGINS OF VIRTUAL REALITY

As a three-dimensional computer graphics capability, VR enables the developer to create pictorial representations, abstract or naturalistic, and display them, with apparent depth, on a computer screen. A VR world can simulate the 'real' conditions of a hospital operating theatre, an aircraft landing, or a racing car in a wind tunnel, and it was in just these sorts of simulator scenarios that the earliest VR technologies began to develop.

The earliest, pre-VR, simulators were devised for training personnel in circumstances where 'real-world' training could prove difficult, expensive or safety-critical. Prior to the development of video or computer graphics, simulators were constructed through a combination of mechanical gadgetry and flat illustrations or photographs. For example, an artillery view-finder could be rigged, via mechanical or electrical connections, to a large painted backdrop, representing the target zone, on which the trainee gunner's successful, or not-so-successful, aim could be displayed by embedded lights. The use of video, as it became available, obviously increased the functionality of such systems, enabling alternative backdrops (perhaps different airports in a commercial flight simulator) to be installed fairly easily. However, computer graphics had two fundamental advantages over this approach: flexibility and interactivity.

2.3 FLEXIBILITY AND INTERACTION

Video or paintings can be used to represent certain limited conditions, but a computer-based model is theoretically unbounded. For example, the same model of an airport may be viewed under a variety of programmed conditions (bad weather, night-time etc.). The scope for designing worlds, and the objects within them, is limitless as real-world constraints, such as gravity, dimension or even common sense, do not have to apply.

Whilst VR can present realistic scenarios, it can also be used to present scenarios that otherwise would be impossible to experience. A developer could construct a rocket for travel to distant galaxies, or reconstruct the streets of ancient Pompeii. Theoretically such boundless opportunity is available in any graphics medium, dependent only on the skill and imagination of the illustrator. The power of VR is that it can take the created world, real or fantastic, and allow a user to interact with it. Interactivity is one of the core elements of VR and separates it from other two- and three-dimensional graphics mediums. VR can allow one or many people to interact with computer-generated objects and worlds in the way that they would interact with the real-world (or other) equivalents. Users can apparently fly to distant galaxies or, if they so wish, stand on the streets of Pompeii just before Vesuvius erupts!

The degree of interaction that users have in a VR world depends loosely on engineering within the world itself and the hardware that they use to interact with it. A VR world is effectively an interface that gives users some feeling of existence within an artificial world created by computer graphics (Vellon *et al.* 1999). Users may be represented in the world in a range of forms: as a complete virtual body (an avatar), as a part of a body such as a hand or as a controllable viewpoint (Shawver 1997). The world can be engineered to give users control of elements within it, for example a vehicle, and navigation can be enhanced by including three-dimensional signposts, instruments or buttons. It is also possible to add text or other two-dimensional graphic aids to a VR world to assist users in their tasks.

A variety of visualisation systems and external hardware devices are used to enable interactions with VR worlds. The level of 'immersion' within a world is dependent upon the devices that are used, and the sort of interactivity that is designed into the world. The most common systems for viewing VR worlds can be summed up as:

- **Projected**. The user's field of vision is effectively filled by screens displaying a projected virtual world. Projection may be onto large concave screens in front of the user or within 'caves' or 'sheds' that users walk into. The latter can fill a 360 degree field of vision
- **Headsets**. Users wear stereoscopic glasses or head-mounted displays (HMDs) which place small screens right in front of their eyes. HMDs enhance users' feeling of immersion/ interaction within a world by excluding any glimpse of the real world and by revising the view of the virtual world as the user moves their head to look around
- **Desk-top**. The virtual world is projected onto the screen of a standard computer monitor. This approach relies on interactive features built into the world to provide a degree of immersion for users
- **Table-top**. The virtual world is projected onto a horizontal table-top screen, and is otherwise similar to the desk-top display. It allows interaction in circumstances where a horizontal format is appropriate. For example, a mechanic could learn how to fix a virtual machine in a way that simulates working on a real table-top.

Specialist hardware devices are available that can give users a greater sense of immersion within the world. These devices include the HMD (Barfield *et al.* 1997) and sensor or data-gloves, which are designed to allow natural movements of the head or the hands in the real world to control movements in a virtual world. However, the standard computer keyboard, mouse, joystick or the more VR-specific spaceball (Jern and Earnshaw 1995) can enable a user to control a vehicle, avatar, tool or viewpoint and offer a level of immersion within a virtual world.

The different levels of immersion within virtual worlds can be defined as:

- **Fully Immersive**. An array of VR specific hardware is used to translate a user's natural movements into virtual activity. Devices include the HMD (described above), sensor or data-gloves and sensors attached to a user's body that detect, and translate, real movement into virtual activity (Cress *et al.* 1997). Devices can also be designed to give users feedback from the virtual world, for example sensations can be stimulated on the skin (e.g. heat or cold) or gloves can physically resist movement when a virtual object is encountered (Luecke and Chai 1997)
- **Partially Immersive**. The hardware that is used in these systems allows users to remain aware of their real-world surroundings rather than being fully immersed in the virtual world. For example, a partially immersive system may include a sensor-glove and a virtual hand but use a desk-top screen for visualisation. In this case, users are fully aware of their surroundings but can interact with the world with natural movements using the glove. Desk-top systems which allow users to control movements using a standard mouse offer a lesser degree of immersion
- **Augmented**. In augmented reality systems, users have access to a combination of VR and real-world attributes by superimposing graphical information over the real-world (Kim *et al.* 1997]. For example, a trainee surgeon could perform an operation on a virtual dummy using HMD or table-top display and a real scalpel. Such a system enables users to develop appropriate motor skills without risk and under a range of different conditions.

2.4 APPLICATIONS

Virtual reality has been described as a 'multidisciplinary effort covering everything from mechanical engineering to psychophysiology' (Rosenblum *et al.* 1994). The briefest of examinations into the applications of VR will support this idea. The potential uses of the technology are boundless, but there are essentially two approaches to current VR development: modelling the real world and abstract visualisation.

2.4.1 Modelling the real world

Some of the more obvious applications of VR are those where a computer permits simulations of the real world in a safe, controlled or more economical environment. The Virtual Reality Annual International Symposium (VRAIS) provides a variety of applications of this type, for example, fire-fighting training, medical examinations, driving instruction, vehicle crash testing or wind-tunnel experiments. This approach also makes the modelling of reality possible in ways that would be intractable in the real world; for example, space or deep sea travel, or even a system for examining social interaction within a family of gorillas (Allison *et al.* 1997). It also permits the sort of model so frequently encountered in TV archaeology, where long-since destroyed buildings are 'rebuilt' and presented in a synthetic environment.

2.4.2 Abstract visualisation

The other most commonly found approach to VR application is in those areas where large quantities of abstract data need to be manipulated, examined or accessed. Such visualisations range from common datasets such as maps, to micro and macro structures such as molecular architecture or social networks. By combining VR with Geographical Information Systems (GIS), geographical information can be explored in three dimensions (Koller *et al.* 1995) or the information contained within a computer database can be visualised and navigated as a solid, tangible entity (Freeman *et al.* 1998; Herman *et al.* 1998; Risch et al. 1996; Teylingen *et al.* 1997).

   Almost any situation that requires interaction with information (even mathematical algorithms (Hibbard 1999)) can benefit from VR visualisation. Users are able to visualise and interact with information through multi-dimensional graphical representations (combined with text clues). Such representations increase users' ability to analyse the underlying data by negating the need for them to construct their own mental image of the data structure (Arndt *et al.* 1995; Serov *et al.* 1998; Stanney and Salvendy 1995).

2.4.3 Distribution

The growth in networked computer systems is also enhancing the variety of VR applications, although the advantages of accessing applications from more than one machine are still being explored. Two areas can already be identified: those where groups of people can interact within a single simulation (see Section 4), and those where information can be disseminated to wider numbers of people. As technologies develop, it will be possible for multiple users to take part in game playing, management meetings or group engineering design in three dimensions across networks (Goodrun 1994; Litynski *et al.* 1997; Macedonia and Noll 1997; Sato *et al.* 1997). Distributing information to ever larger audiences, or at least making it available to the potential audience, is useful in almost every field. Communicating scientific findings and access to public information have already been enhanced through the use of distributed systems (Rowley 1998).

As the World Wide Web and its associated technologies develop, there will doubtless be an increase in the applicability of presentation mediums such as VR. There will also be a corresponding growth in the number of different applications (Encarnacao and Fruhauf 1994; Fuurht *et al.* 1998).

2.5 VIRTUAL REALITY FORMATS

As the number of applications of virtual reality (VR) has grown, there have also been changes in the different formats of VR-type software. Each format has differing approaches to, and varying degrees of, three-dimensionality, immersion and interaction. Whatever the format, as a result of the need to provide clear and fluid imaging that constantly changes as users move within the world, VR requires substantial processing power. Until relatively recently, VR systems were restricted to very expensive graphics workstations. Increasingly though, VR is being exploited on personal computer (PC) platforms as a result of their increasing processing power and improvements in graphics delivery hardware and developments in PC-based VR software packages and formats.

The benefits of these developments can be seen on the Internet where they enable increased activity in three-dimensionality and interactive graphics development. VR presence on the Internet is currently dominated by the Virtual Reality Modeling Language (VRML) standard. This language has been developed to provide a multi-platform, universal language for interactive three-dimensional graphics across the Web (Carson *et al.* 1999; Nadeau 1999). Many applications already exist which utilise the benefits of VR in the VRML format (Earnshaw 1997), but the format has not grown, or been accepted, quite as quickly as originally thought. This is, in part, due to the limiting factors of cross-platform, cross-browser compatibility, and also to less-than-perfect technical and political development. Various bodies have attempted to develop VRML standards, including Microsoft, Sun Microsystems and Apple, but all were rejected in favour of VRML 2 which has since moved onto VRML 97 (see Section 3.7).

VRML is not the only PC-based, web-compliant VR format. For example, Superscape created its own format (SVR) which ran efficiently and effectively, through either Netscape or Internet Explorer, via its own viewer Viscape. A number of SVR models can be found on the Web covering a variety of applications (including entertainment, marketing, training, and data visualisation). However, Superscape no longer retails its VR-authoring software, the company now focuses its business on 3-D applications for wireless devices (Superscape 2002).

The withdrawal of VR products by manufacturers (see Section 3.10), combined with the apparent slow acceptance of VRML as a securely established medium, should perhaps provide something of a warning to all developers of VR models or applications.

2.6 THEORETICAL CONSIDERATIONS FOR THE DEVELOPMENT OF VIRTUAL REALITY PROJECTS

The previous sections have outlined the technical aspects of VR, and described some of the application areas in which the technology has been successful. This section looks at the reasoning behind VR development: those areas that must be considered by any developer of a VR project. These considerations can be divided into two issues: one relating to the need for appropriate level and style of content; and the other relating to the need to use the right software and delivery mechanisms. In essence, it is imperative that a developer considers both the information and the technology in any IT-based project. Both of these issues have a common thread, in that identifying the appropriate constituents of either requires an understanding of the intended user.

2.6.1 Data to information

The key to the successful delivery of any VR model is that it addresses not only the needs of the developer, but also truly matches the needs or desires of the user who will be at its receiving end. This may seem obvious, but it is remarkably easy to become embroiled with the 'back-end' of a development; to be side-tracked by technological issues or shrouded by personal beliefs and developer-centred perspectives. This can be avoided by conducting a formal analysis of user requirements or, if the project is of a smaller scale, at the very least initiating some concerted research or debate. What all computer-based systems are ultimately trying to achieve is to impart, to varying degrees, a level of information to a user. Virtual reality aims to take a collection of otherwise unintelligible data (points, facets etc.) and present it on screen in a way that the user can interpret it, in keeping with the developer's hopes or intentions. This is true whether the information is a representation of complex numerical data aimed at scientists, or a moving abstract image intended to stimulate a subjective, personal response. For the imparted information to achieve its desired goal, understanding by the user is imperative. The process of turning data into information is illustrated in Figure 1.

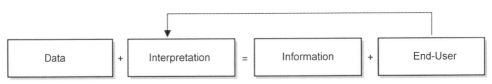

Figure 1: The simplified information process

VR images, like any other graphical illustrations, are merely vehicles for elucidating, or clarifying, information to the user. It is clarification, not realism or accuracy, that is at the centre of any illustration, and it is critical to consider this when new technologies or methods, for illustration of any kind, are being developed. The main goal of an illustration must be to fulfil the needs of its intended user; to clarify the information in a manner, and at a level of abstraction, appropriate to the intended audience.

2.6.2 Graphical information

Interpretation of data to provide information will always raise subjective issues. The very notion that information should be placed into a context that is 'appropriate' for a user confirms that different people will tend to interpret the same thing in different ways. This point of view is particularly important when dealing with images: everybody knows that a picture paints a thousand words; the problem is that they can often be a different thousand words for each viewer! Computer graphics, and VR in particular, convey a substantial quantity of information to the viewer, and obviously can be a very powerful component of any modern computer-based system.

There are drawbacks as well as advantages. The plus side is fairly obvious in its nature, but quite surprising in its potency: many people are impressed by flashy, animated computer graphics. As Brooks advocates, graphics relate much more directly to the way humans naturally interpret and communicate information (Brooks 1996), and this affects the way people respond to a highly graphical product. This is very useful for improving end-user perception, because if the user finds interacting with the system a pleasurable, or fun, experience, then navigation and information-gathering are more likely to be successful. The perceived impressiveness of naturally interactive and visually stimulating graphics can win over all but the most hardened heart.

VR modelling, with full three-dimensionality, extensive animation, and elaborate fly-throughs can provide something of a cost problem in terms of both time and money and also in terms of download times and user-system requirements. These are issues that any such development must face. However, it is the user requirements against which these risks should be assessed, even when trying to resolve problems relating to the technology.

2.6.3 Technology for informing

Often, much of the time and money cost of a VR-based product or system can be attributed to the technology, yet the solution to cost-cutting (of time, money and effort) is invariably found in the manipulation of the inherent information. It is too easy to become embroiled in the belief that the only way to solve a computer-related problem is to have the latest in software, hardware or expertise. This approach can be likened to the idea that a book can only have value if it is printed using the most sophisticated techniques, in the smartest fonts and on the most expensive paper. What is important is that the product reaches its intended end-user. Many potential users of VR models will not have access to the same hardware and software as the developer.

The artificial worlds that people experience in VR are all created by a developer, designer or artist through the manipulation of data and information which is then represented on screen. VR developers can be lured by the impressive capabilities of modern graphics software and hardware (the stunning results of which can be seen in films such as *Jurassic Park*) into the belief that 'reality' should be gauged by closeness of the model, in appearance and movement, to its real-world counterpart. However, VR is essentially a graphics medium and it should be remembered that good illustration, or graphics, or art, depends not on how 'realistic' an image is, but on how successfully it conveys its intended message. For example, an entire city can be abstracted down to the level of a black dot on a map. This is sufficient detail for a map of the world but useless for someone requiring a town street-guide. Suitable abstraction of this kind is a necessity because, as Miller and Richards point out, it would be impractical for a map to show *all* the data (Miller and Richards 1995). Not only impractical, but very expensive.

Another important element to be considered is the mechanism for delivering the VR to the end-user, i.e. the software and hardware that users will require. Understanding the technical limitations of the model's intended users will be part of the formal analysis of user requirements (see 2.6.1 above) if it is being designed with users in mind. There is no point in developing a VR product if the intended users will never see it! Where the project involves delivery via on-site monitors this is less of an issue, as the developer has full control over the delivery mechanisms. However, where the project involves delivery via the Internet, developers do not have this advantage. Most VR products, including even the simpler QTVR type images, require users to download and install a plug-in to view the model. It is tempting, as a developer, to believe that this will not be a problem for users and that most will not begrudge the time and trouble involved.

There is more to it than that. Most importantly, users do not *know* that the VR will provide delights and may well *not* be prepared to expend too much energy to find out. Many links on the Internet make promises that are not fulfilled. Furthermore, the model itself will also need to be downloaded, the plug-in may not be compatible with the browser version or the operating system that are in use and the graphics capabilities or settings of the user's PC may render the image incorrectly or not at all.

These problems are not insurmountable, providing enough research is done initially to find out who is likely to use the VR. With research, it is possible to ensure that the VR's technical requirements conform to capabilities of 'standard' users, even if it is not totally inclusive. Understanding and consideration of the user is the key to any successful project delivery.

Section 3: Virtual Reality Methods and Techniques

3.1 INTRODUCTION

There are several formats for developing desk-top virtual reality and new formats are continually emerging. For those considering creating virtual reality, the format chosen will depend on a number of factors. The first considerations will be whether the format supports the features required and how successfully these can be delivered to the intended audience. It is important to consider whether the preferred development software is available for the platforms used both during development and by users. This includes considering whether plug-ins or viewers are needed to view the virtual reality, at what (if any) cost and whether these run on the users' likely range of computer platforms. When selecting a format, it is important to consider the lifespan of the virtual reality world to be created and the implications of maintaining and archiving it throughout that period. Other considerations are the investment needed to buy software and hardware and the time required to learn new skills and create the worlds, all of which vary from one format to another.

This section looks at general practices relating to the creation of virtual reality worlds and then considers web-deliverable virtual reality formats: bubble worlds (including QTVR), VRML, Java 3D and X3D. Although bubble worlds are not true virtual reality, they are included for the purposes of this Guide because they give an experience close to virtual reality and limited investment of time and resources is needed to create them. Each format will be introduced and consideration given to issues relating to viewing the worlds, their functionality, and the skills and equipment required to develop them. Advice is also given on how to get more information.

Although this Guide includes references to specific software products, this should not be taken as a recommendation from the AHDS. Similarly, the order in which the software is introduced has no significance.

3.2 GOOD PRACTICES FOR CREATING VIRTUAL REALITY

3.2.1 Why use virtual reality?

Perhaps the first phase in a successful virtual reality project is deciding whether it is in fact the correct solution to your requirements. Why are you choosing virtual reality? 'Because it is there' may not be a good enough reason. For instance, in archaeological reconstruction, the use of virtual reality might be inappropriate if it misleads, suggests a certainty that does not exist or if it inhibits reinterpretation by suggesting that a reconstruction is carved in stone. Virtual reality may not be suitable if the target audience does not have access to the technology needed to view it or lacks the computer skills necessary to manipulate the model. Always consider the return that will be gained from the use of virtual reality. If the return is not significantly higher than that gained from using traditional techniques then the extra investment needed to create VR may not be worthwhile. In saying this, a well-planned, well-documented and well-executed VR world can impart complex three-dimensional information much more successfully than traditional media.

Virtual reality as a solution should be considered only if its use will assist in your objectives efficiently and in a manner which is superior to other technologies. Having chosen to use virtual reality, creating a successful world requires a systematic approach.

3.2.2 Project definition

Like any other media project, you need to begin by preparing a storyboard or a good description of the world that you want to create. This will outline on paper your ideas for the world, the features that it will contain and the type of interaction that it will offer to users. In addition to the storyboard you also need to prepare:

- An analysis of user requirements
- A definition of the objects, textures, animations and sounds that will make up the world
- Documentation for the design of the model and how it works within the chosen system
- A definition of the platforms which your world will run on
- A test and maintenance schedule, including usability testing
- A projection of the anticipated life span of the virtual reality model.

For an example of a project brief, see the Case Study Library for Narrative Rooms' description of the process that they went through in preparing the on-line exhibition 'Brancusi's Mademoiselle Pogany' for Philadelphia Museums.

3.3 USER REQUIREMENTS

Once a design specification and storyboard have been prepared for a virtual reality project, the next step is to think about how users will interact with the world and about their expectations from the world. Some of these considerations are discussed below.

3.3.1 Ease of use

The length of time that users are likely to be willing to invest to use a world generally reflects the length of time that they spend viewing it. For example, if a virtual world is displayed in a museum then users generally will use it once for a short period of time. If the world is intended for use at home or in a school or college, users will generally have more time to view it and may be willing to invest more time to learn how to use it. Consideration also has to be given to the complexity of the world and to the fact that certain types of users will have more advanced technical skills than others. This can change depending on where the model is accessed.

3.3.2 Conventions in navigation

Users may have preconceptions about how navigation or interaction with the world will work. These may be based on previous experience or on intuition. Developers need to make sure that the user interface follows standard conventions where these exist. For example, many people are familiar with Windows conventions to exit a program and some will also be familiar with the standard conventions used for navigation in a particular browser plug-in. Developers are recommended to adopt any existing conventions and try to make local conventions compatible in style so that they are easier for users to learn. Navigation should be obvious and intuitive so that users know how to get around the model easily and can return to the start or go back should they want to.

3.3.3 Realism

As more virtual worlds are seen in computer games and on the television, users have growing expectations about their level of realism. Developers need to consider what level of realism users will be satisfied with while at the same time not suggesting a level of certainty where none exists. Documentation should be kept when constructing the model so that at the end of the building process it is easy to tell which parts are based on evidence and which are works of imagination (see Section 5.1). This should be conveyed to the user in some way, perhaps through an explanation of how the model was built.

3.3.4 Degree of interaction and movement

The extent to which users wish to interact with virtual worlds varies with different audiences. Children may be bored if they can't interact while non computer-literate adults may be intimidated if they have to navigate by themselves. Pre-set animations and fly-throughs may be appropriate for one audience but not for another.

3.3.5 Method of interaction

Users can interact with virtual worlds by using a mouse, keyboard, joystick or a touch-sensitive screen. Different methods of interaction are appropriate for different situations. For example, a touch-screen offers a robust user interface suitable for a museum display. Users with disabilities may have very specific requirements for interacting with a virtual world. See the case study 'Exorcising the Flesh' by Kate Allen for an exploration of people's interactions with her Fine Arts VRML installations.

3.3.6 Speed

Users have different expectations about how quickly they should be able to move through a world or how long they will wait for a response. Ideally, users should have a sense of smooth movement and should not have to wait so long that they begin to doubt whether things are working properly. Speed of use is directly related to the hardware and software platforms being used to view the world. In some cases the speed of the network connection also needs to be taken into account.

3.3.7 Hardware and software

Different audiences may view your world from different hardware and software platforms. For example, home users are likely to access your world using a desk-top computer and a dial-up modem. Developers who are creating virtual worlds for home users may need to compromise between satisfying the user's expectations of the feel and visual quality of the model and file size (see Section 3.5).

Another important consideration which may influence the choice of development software is whether plug-ins are needed to use virtual worlds. Different audiences may be more or less willing to access and download a plug-in depending on the length of time that they anticipate using a world. For example, users of an educational world are probably more willing to invest time in such installations. Home users might be reluctant to install plug-ins and may prefer to use models that either do not require them or use operating systems that incorporate plug-ins as standard. Plug-ins may also be difficult to obtain, become obsolete or not work with future versions of the platform.

See Brancusi's Mademoiselle Pogany for a discussion of how user requirements influenced the delivery of the virtual reality in this project.

3.4 BUILDING THE WORLD

The process of preparing a storyboard involves defining the elements that are required to create the impression of objects and spaces in a virtual reality model and to enable users to interact with a world. For example, users of a model may be able to open the door of a building and to hear a welcome message. This simple interaction is made up of a number of separate elements which must be defined in the project storyboard because of the implications of creating them in terms of time and resources. This section of the Guide will introduce the objects, textures, sounds, animations and program scripts that make up a virtual world.

3.4.1 Objects

The objects that will make up your world can be created using authoring tools, CAD software, 3-D scanners or by stitching images together. The techniques that are used to produce objects will depend on the software platform in which the VR is being developed. For example, in Bubble Worlds (see Section 3.6) panoramic images are stitched together to create cylinders or spheres giving the impression of a three-dimensional space or object. In VRML (see Section 3.7), polygons are combined to create three-dimensional objects in the virtual world.

Authoring tools

Polygonal objects can be authored directly; for example VRML developers can use a text editor to define the facets and surfaces that make up an object in the modelling language. Another option is to use a VR authoring tool such as a polygonal modeller or 3-D authoring software (e.g. AutoCAD, 3D Studio Max) to ease the process. Polygonal modellers allow users to create objects from scratch or to edit objects that have been acquired from another source, such as an object library. Drafting software can be used either to draw an object from scratch or to create a model from measurements taken in the field, e.g. from a survey of a real-world building. CAD packages are often used to create 3-D models (see Eiteljorg *et al.* 2002) which are exported as VRML files or sometimes imported into VRML using conversion software.

The number of facets that make up an object has a direct impact on the speed with which the object is drawn on screen and also on file size. The more detailed an object is (i.e. the more facets it has) the longer it will take to draw and the larger the file size. Some polygonal modellers allow exact control over each facet that makes up an object. Such control helps to create an efficient world but editing each facet can be very time consuming. In these terms an efficient world is one that has struck an appropriate balance between detail and rendering speed for the platforms on which it will be used.

Optimisation tools are also available which improve efficiency by removing any unnecessary facets that slow down the rendering of the object. Some polygonal modellers automatically create less detailed versions of objects by reducing the number of facets. This is useful for Level of Detail (LOD) operations where less detailed versions of objects are successively replaced as the user's viewpoint recedes to set distances. LOD operations prevent the computer from rendering detailed objects that the user cannot 'see' from his or her viewpoint and help to speed up the rendering of the world.

3-D object scanners

3-D object scanners use either a laser or white light to capture the surface of an object as a point-cloud of co-ordinates. White light scanners also capture a colour texture map of the object. A range of different types of 3-D scanners is available. Some are designed to work at close range and are suitable for small objects while others work at ranges of up to 1000m and are suitable for use on buildings. Scanners remain very expensive at present.

Bubble worlds or panoramas

A bubble world is a seamless panorama which is created by stitching or blending one or more images together using a software application. The application joins the images together to create the illusion of a cylinder or a sphere and removes any obvious join lines between images. Some packages allow for the entire panorama to be exported to a third-party photographic manipulation package for editing.

Object libraries

Objects can be obtained from object libraries, for example the Web3-D Repository. Some objects are free, others are available for purchase and a wide range of different objects is available. Such objects can be combined with others that have been obtained from different libraries or ones that you have created using an authoring tool. Objects that have been created in one software package may be used in another – for example objects created in CAD are often converted to VRML. When obtaining such objects it is important to check:

- That the objects have been saved in a format that can be read
- The units of measurement in which the objects have been created. Objects can be scaled within the VR, but they may then lack detail or be overly complex for the scale at which they are viewed. Without scaling, the objects may appear to be either larger or smaller than they should be in relationship to other objects within the world
- Whether objects have been optimised (unnecessary facets have been removed) so that they are rendered quickly in real-time applications

- That any objects taken from an object library do not have copyright restrictions relating to them.

3.4.2 Textures

Texturing is a technique which involves applying images that represent surfaces, building materials or other surface details on 3-D models. It is an effective way of making virtual worlds appear more realistic. The textures can be created by taking photographs of real objects or by creating digital images. Each texture is stored as a bitmap or other image format (depending on the system used) and requires much more file space and rendering time than simply assigning a colour to a facet of an object. There are several ways of minimising file size and the time taken to draw objects:

- Repeating a simple image over a facet rather than using a single image for the whole facet; this is called 'tiling'. For example the appearance of a brick wall may be created either by repeating an image of a row of bricks or using an image of the whole wall
- Creating textures at different levels of detail for use in LOD operations
- Creating texture bitmaps at the lowest resolution possible
- Minimising the number of facets used for an object
- Providing texture-free versions of the world for users with slower modems
- Using compression on image files to create smaller versions, e.g a jpeg version of a bmp file.

3.4.3 Sound

A wide range of tools is available for creating and capturing digital sound files for inclusion in VR worlds. Sounds and sound effects are also available from sound libraries.

3.4.4 Animation and scripts

Program scripts can be attached to the objects in your virtual world to animate them or to control what happens when users interact with them. Animations involve the movement of an object or viewpoint along a path that has been predefined in a script, for example in a fly-through. They often take place in response to a user action, for example opening a door, and may be repeated in a continuous loop or have a set start and finish point. Examples of animation and scripting include:

- A user clicks on a door and the door opens or closes. The program script determines whether the open or close animation is run and controls the angle by which the door object rotates as it opens or closes
- A user clicks on an avatar and hears a welcome message and sees the avatar's mouth opening and closing
- When the user drags the mouse forwards, the viewpoint moves forwards until the user stops
- If a user bumps into a wall then their movement stops
- A fly-through or guided tour is repeated continuously as long as the world is active.

The combination of animation and scripted behaviour leads to worlds possessing rich levels of interactivity. Scripting requires programming skills and further development time. Different scripts can be used – Java and ECMAScript are examples but any programming language that browsers support will work. The language used must be carefully chosen if models are to be used on different browser platforms.

3.4.5 Finding out more

See the case study on Virtual Saltburn by the Sea by Clive Fencott for a discussion of the ways that textures, sounds and animations can be used in virtual worlds to shape user interactions.

The Web 3-D Repository (http://www.VRML.org/vrml/vrml.htm) maintains lists of authoring applications, software development resources and libraries of objects, sounds and textures.

3.5 HARDWARE, TESTING AND MAINTENANCE

3.5.1 Testing for target platform(s)

The hardware and software platforms used to create your world are unlikely to be the same as the platforms used by your target audience. The likely platforms should have been identified during the planning stage of your project (see sub-section 3.3.7). Once the world has been developed, it is important to view it on all of the target platforms, check that it achieves the desired performance and make any modifications. You will need to consider:

*   CPU speed
*   Browser configuration
*   Internet connection (or other delivery options such as CD-ROM)
*   File size
*   Visual quality

The visual quality of the world, together with the sounds and animations that it incorporates, have an impact on both the overall file size and the rate at which the world is delivered on screen.

The frame-rate is the number of times per second that the world is redrawn. For the human eye to perceive a smoothly changing view it must be redrawn at a rate of at least 25 frames per second. A computer's ability to redraw a scene at this speed is determined by the number of facets that have to be displayed at any particular viewpoint and by the textures, animations and other processes being performed. The complexity of information that must be redrawn for a given viewpoint can be minimised by level of detail (LOD) techniques. This involves displaying less detailed versions of objects at greater distances from the user's viewpoint. However, this technique involves creating several versions of each object to be displayed at different viewpoints and thus increases the overall file size (although speeds up the rendering time of individual sections of the model).

File size and the types of Internet connections being used are important considerations if the world is to be downloaded from the web. How long should a user be expected to wait for a world to download? A user in a university may have an Internet connection that allows files to be downloaded up to a rate of 10 megabits a second (although the actual rate depends on the number of users on the university network at the time). However, home users will be able to download files at much slower rates – dial-up modems download files at 28 or 56 kilobits per second while domestic DSL lines achieve rates of up to 512 kilobits per second (1 megabit = 1,024 kilobits). Thus home users will wait much longer than university users for files to download. File size is determined by the number of objects, textures and sounds within the world. It can be reduced by using a compressed format for web delivery, if one is available for the type of virtual reality you are creating. Image files, such as those used for tiling and texturing, can usually be compressed using the normal techniques and experimentation will show the effect that this has on the model.

Another important consideration is that virtual worlds are usually developed on quite fast, powerful computers. Users may not have the same capacity when viewing the worlds, which is one of the reasons why it is important to test on other machines.

During testing it is important to consider the balance between the visual quality of your world, the frame-rate, file size and their impact on the delivery of the world to users. Consider whether realism is more important than speed or if it should be the other way around.

3.5.2 Maintenance and testing

Web-deliverable virtual reality is almost always reliant on a plug-in or viewer running in conjunction with a browser. As new versions of these are released, it is important to test that your worlds continue to run as intended. There may be differences in the way that browsers implement scripting languages (such as Javascript) used in virtual worlds. You will need to test (and continue to test) that worlds perform in the same way within all common browsers. There is also the general issue of maintenance of the system – who will maintain and update the model after the project ends? Who will be available to deal with any problems when the model is in use? This is one of the reasons that full documentation is necessary (see Section 5).

3.6 BUBBLE WORLDS

The term 'bubble worlds' covers a large range of application packages designed to offer a quick, easy and cheap way to present landscapes and indoor environments in the round. They are a simple technique for giving a general impression of a place and are excellent for guided walks.

Bubble worlds can be created with standard camera and computer equipment and with no special training, although special hardware is available to assist. All application packages take a broadly similar approach. The technique comprises the generation of a seamless panoramic image that is projected onto the inside surface of a 'notional' cylinder or sphere and then viewed through an interactive window on the computer screen. This can give the impression of viewing an entire space from the ground to the sky and 360 degrees around through a moveable window.

Bubble worlds were designed for Internet delivery and can be viewed using Java-enabled browsers or browser plug-ins. As a result they are becoming a favoured technique for presenting snapshots of full virtual reality worlds on the Internet.

3.6.1 How to view bubble worlds

No special hardware is required for viewing bubble worlds. They can be viewed on low-specification computers, normally via a web browser. Most bubble world formats depend on either Java applets (which require a Java-enabled browser) or browser plug-ins. Some bubble world formats require a stand alone application that is invoked by a browser.

The navigation tools vary for each bubble world format; as a minimum users will be able to pan, tilt and zoom. It is the ability to pan, tilt and zoom via the viewing window that gives the impression of three-dimensionality so apparent with these products. Bubble world interfaces tend to be intuitive and easy to learn in comparison with the more complex interfaces of VRML clients.

3.6.2 What is needed to create bubble worlds?

This is not a standardised technique and a wide range of software applications is available to create bubble worlds. However, Apple's Quick Time Virtual Reality (QTVR) is becoming a *de facto* standard for bubble worlds because of its ease of use, reliability and interface design. QTVR offers file formats for both panoramas and object movies, a browser plug-in, stand-alone viewer and a comprehensive range of authoring tools.

Bubble worlds can be generated from a sequence of overlapping photographs taken from a single point. Special lens attachments allow full environments to be captured in a single image, although image quality is compromised (there is inherently more information in a sequence of images of any given scene than in a single image of that scene at the same resolution). Photographic prints can be scanned for manipulation by the production packages but use of a digital camera greatly simplifies the procedure. The ultimate delivery medium and intended use of the final panorama must be considered when capturing an image. For example, a single-image panorama is quick and simple to take with a specially designed lens and produces a bubble world perfectly suited for a website but unsuitable for projection on a large screen during a lecture. Users may prefer to download a higher resolution bubble world to run locally rather than run a lower resolution version on-line. Project managers may choose to capture high-resolution images and/or panoramas for archive purposes while delivering a lower resolution version of the world on-line; this approach allows for upgrading of the delivered product as hardware and software specifications improve.

Some production packages are very tolerant and adequate panoramas can be generated by standing in an environment and taking the photographic sequence by hand. However, this is not recommended, and expensive and time-consuming trips can be avoided by capturing the highest quality sequence of images on the first occasion. The best way of doing this is to use a panoramic head, a tripod attachment that allows the camera's recording plate to sit at 90 degrees to the horizontal plane, centred over the point of rotation. This minimises distortion when separate images are joined or 'stitched' to create a single panoramic image. Panoramic heads can be used with the camera in 'portrait mode' allowing greater vertical coverage for each image in the sequence. Although more shots are required to capture the full 360 degrees, a separate sequence is not required to capture higher or lower sections of the environment.

Not all panoramas need to be 360 degrees. If file size is an issue then a 180 degree panorama at higher resolution may be better than a 360 degree one at lower resolution. For most landscape panoramas a single sequence of images in portrait orientation will be adequate.

Panoramic heads often have graded positions that allow the camera to be rotated at exact angular increments. This results in better quality panoramas as most software packages assume that each image covers a specific spread of degrees and that the overlap between images is constant.

Once the image or image sequence has been captured the images are 'stitched together' then 'blended' to remove any obvious join lines. At this stage some packages are better than others and, depending on the initial camera settings and lighting conditions, it may be very difficult to get two separate images to blend together without the 'join' being obvious. The better packages allow for the entire panorama, or panoramic tile, to be exported to a third party photographic manipulation package for editing. This eases the process of removing unwanted elements, adding reconstructed or speculative elements and manually blending and smoothing joins between images.

Factors affecting the presentation of a bubble world can normally be edited when the panorama is created. For example the window size, starting position, zoom range and file compression can be specified. Several bubble worlds can be linked together by hyperlink 'hot spots', generating a series of navigable scenes in a single file. Users can move from point to point within the environment by moving from one panorama to another. In web browsers other hyperlinks can move the user to text documents or still images. Some production packages allow various effects such as sound and streaming video to be embedded in certain types of bubble world.

3.6.3 What are object movies?

QTVR allows for the creation of object movies. If a panorama can be imagined as being projected onto the inside of a 'notional' cylinder then an object movie is the reverse of this. The images are projected on the outside of the cylinder. A sequence of images of an object are used to create the impression of three-dimensionality but much more quickly, cheaply and easily than by actually generating three-dimensional models. As with bubble worlds, a range of additional functionality can be added to object movies. As well as allowing objects to be rotated, imaginative use of the technology allows for other conditions to be changed interactively, such as lighting or background.

3.6.4 How to find out more

A useful starting point for more information is the Panoguide website maintained by James Rigg. This site includes lists of software, costs, tabulated summaries of features, reviews of performance and is kept up to date as new software appears. See http://www.panoguide.com.

See the article by Stuart Jeffrey in *Internet Archaeology* (Jeffrey 2001) for a discussion of the uses of various techniques for visualising three-dimensional models of sculptured stones in their landscape context.

3.7 VRML: VIRTUAL REALITY MODELING LANGUAGE

3.7.1 What is VRML?

Virtual Reality Modeling Language (VRML) was developed by the Web3D Consortium and was designed for use on the Internet. VRML is both a scene description language and a file format for virtual worlds. The language is used to describe the geometry and behaviour of three-dimensional scenes. VRML is a popular 3-D format for the web because of its relative ease of use, standardisation and the comparatively small file sizes that it produces.

Three specifications for VRML have been developed. The Web3D Consortium defined VRML 1.0 as a minimum specification to get VRML off the ground quickly and then continued development work, adding features and higher levels of interactivity and releasing VRML 2.0 as an International Organisation for Standardization (ISO) Committee Draft in 1996. VRML 2.0 has now been superseded by VRML 97, which has been approved by the ISO and published as International Standard ISO/IEC 14772-1:1997. There are a number of VRML 2.0 worlds on the Internet; however, there is backwards compatibility between VRML 2.0 and VRML 97 and these worlds can be used successfully in VRML 97 viewers.

VRML allows for the description of hierarchies of simple shapes such as cubes, cylinders and spheres. More complex shapes can also be defined, as can surface materials, texturing of facets and level of detail (LOD) data. Objects can be linked to other URLs via hotspots. Other features include transformation (the reuse of objects more than once), viewpoint setting (which allows users to look at pre-defined views of the world), the definition of lighting within the world and 'shapehints' (which define how particular objects and object types will be rendered). VRML also works with other standard file formats in use on the Internet. Sounds, textures and animations can be linked to objects described in a VRML file by referring to image files, sound clips and program scripts in standard formats.

Features that were first introduced in VRML 2.0 were made easier to implement in VRML 97. The most important of these features are 'sensors' which perform collision and proximity detection. Checks can be built into the world to test the visibility of an object from a given viewpoint. User interactions such as clicking and dragging are allowed and time can be measured and expressed. Importantly, there is the ability to incorporate scripts which program behaviours in the worlds, for example scripts can be used to change the lighting as a user moves around a space (see Case Study 3 for more examples of programmed behaviours).

3.7.2 How to view VRML worlds

VRML viewers are needed to allow users to navigate their way through and interact with VRML worlds. Most web-browsers now include a VRML viewer. However, some browsers do not and users must download a plug-in or viewer from the World Wide Web. A number of different viewers are available for download and many are free. They differ in the style of navigation and performance that they can offer users. VRML developers often point users to a particular viewer. Users, on the other hand, may select a different viewer because they are familiar with its style of navigation.

The Cosmo Player is one of the most popular viewers because it offers a wide range of movements. The Cosmo Player runs on a wide range of operating systems and is now distributed by Computer Associates International. VRML browsers that support multi-user shared worlds are also available, for example Blaxxun Contact 4.0 (see Section 4.6).

A list of all of the browsers that are currently available can be found at the VRML repository, http://www.web3d.org/vrml/vrml.htm.

3.7.3 What is needed to develop VRML worlds?

VRML can be written using a text editor and viewed through a web browser, requiring no financial investment other than the time taken to learn how to write VRML and access to a computer. There are many books available on writing VRML and there are also some good manuals and on-line tutorials on the World Wide Web. However, hand coding is time consuming, can be tedious and it can be difficult to spot problems and debug the resulting code.

Another option is to use a VRML world-building tool. These packages allow authors to define worlds graphically and save them as VRML. This process is much faster and easier than hand coding but is more expensive. Often CAD packages are used to create 3-D models which are then exported as VRML files; CAD to VRML converters may also be used. Textures, sound, interactivity and behaviours are then added to the VRML using a text editor. There are dozens of world builders available, varying in price and quality, with new systems emerging all the time. The VRML repository includes a list of world builders along with syntax checkers and optimisation tools. When selecting world builders, developers are recommended to select those which will produce VRML that complies with the ISO standard.

Once worlds have been created, a syntax checker can be used to check that the VRML code is correct. Optimisation programs can also be used to improve the performance of the world by removing redundant shapes from the code.

Large libraries of VRML objects, textures and sounds are available on the World Wide Web. Some libraries are copyright free, others require the copyright to be credited and other libraries operate on a commercial basis. See Section 3.4 for a discussion of issues to consider when selecting items from libraries. When selecting objects it is important to check the units of measurement that have been used and whether they have been optimised so that they are drawn quickly on screen.

3.7.4 Problems with VRML

Although VRML is the most popular format available for delivery of 3-D models on the Internet, there are some problems associated with it. VRML is not the best virtual reality system. It is rather a 'jack-of-all-trades', providing many basic functions which are designed to run on all platforms. It can never replace more sophisticated specialised VR systems optimised for specific tasks or configured to run on specialised hardware. Although standardisation helps developers to make sure that VRML files will be delivered consistently on different browsers and plug-ins there are differences, particularly in their handling of lighting and colour. However, the browser manufacturers are co-operating to improve conformance and as a result of their work, together with that by Eric Haines on the implementation of the VRML colour and lighting model, the differences are becoming more subtle.

At the present time, VRML is the best standard that exists for publishing, constructing and viewing virtual worlds on desk-top computers. VRML is currently the best standard for archiving and reuse of virtual reality, although the Web3D consortium are working on developing X3D (see Section 3.8) as the successor to VRML.

3.7.5 How to find out more about VRML

The first place to go to find out more about VRML is the WEB3D consortium at http://www.web3d.org and the VRML repository at http://www.web3d.org/vrml/vrml.htm. Both of these sites are kept up-to-date and contain many useful links. See these for specifications, user guides, tutorials and also for lists of world-building tools, converters, syntax checkers and optimisation tools.

The VRML 97 specification International Standard ISO/IEC 14772-1 is available for download from http://www.web3d.org/fs_technicalinfo.htm.

For a very accessible introduction, see Bob Crispin's VRML Works at http://hiwaay.net/~crispen/vrml/. This site guides you from the process of choosing a browser to writing your own worlds.

3.8 X3D

3.8.1 What is X3D?

X3D (eXtensible 3D) is being developed by the Web3D consortium as the next-generation open standard for 3-D on the web and as a replacement for VRML. X3D has an extensible format which uses XML (eXtensible Mark-up Language) to express the geometry and behavioural capabilities of VRML (see Section 3.7). In developing X3D the aim has been to address the limitations of VRML and to provide a fully specified standard.

The development of the standard has involved a browser working group (including browser companies such as Blaxxun, Nexternet, OpenWorlds and ParallelGraphics) and a task group. The aim has been to develop a standard which can be supported by content creation tools, proprietary browsers and other 3-D applications, and which is also compatible with existing VRML models as well as with the MPEG-4 and XML standards. X3D was launched by the Web3D consortium in August 2001 who made the specification available for implementation and review with the longer-term aim to submit it to ISO (the International Standards Organisation) for ratification. Since then the X3D specification has been actively developed, with the Final Committee Draft (SO/IEC FCD 19775:200x) being submitted to ISO for review in December 2002.

The X3D specification includes backwards-compatible VRML and XML encoding, a Scene Authoring Interface, components to provide extensibility and a lightweight profile that is recognised by MPEG-4. There is also a register of extensions to X3D, including details of their implementation.

3.8.2 Viewing X3D

The X3D browser working group and browser companies have been working to develop viewers which can be used to render both X3D and VRML content. For example, the Java 3D working group (see Section 3.9) and Sun Microsystems have been developing an open source X3D browser called Xj3D. Blaxxun intend to develop their open source browser. Contact VRML 97 browser for use with X3D content.

3.8.3 Authoring tools

Some existing modelling and animation packages already export to formats that are compatible with X3D. For example, Open Worlds is an X3D compatible system that was launched in August 2001 and provides a software developer's kit for 3D multimedia graphics applications. X3D-Edit is an integrated XML authoring tool which allows for VRML to X3D integration and is available from the Web 3D consortium.

3.8.4 Finding out more

Web3D consortium – X3D specification (http://www.web3d.org/fs_specifications.htm)
Web 3D repository – X3D resources including the Open Worlds Horizon Browser, X3D-Edit and XJ3D (http://www.web3d.org/vrml/x3d.htm)

3.9 JAVA AND JAVA 3D

3.9.1 What is Java?

Java is an object-oriented programming language developed by Sun Microsystems (Sun) designed to be portable across multiple platforms. It achieves this portability by using the 'Java virtual machine' (nothing to do with virtual reality), the core component of the Java Runtime Environment (JRE). Software developers compile Java programs to be executed by the JRE rather than a specific operating system, which is the case for applications written in other programming languages. Thus Java developers rely on Sun Microsystems to create or license-out a virtual machine that will port the software across a number of operating systems.

The decision to include a JRE in the Netscape and Internet Explorer web-browsers was a major contributing factor in the rise in popularity of Java for use on the Internet. A web-browser that has a JRE incorporated into its program is known as a Java-enabled browser.

Unfortunately, Sun's ability to extend the capabilities of the Java language has outpaced the browser developers' willingness to update the JRE in their software. Since its introduction to the web, Sun has released several major upgrades to Java but Netscape and Internet Explorer only upgraded to Java version 1.1. The 'classic' Mac operating system includes the original version of Java and has never been upgraded, although Apple's latest operating system, MacOS X, now includes Java 2. To remedy the problem of Java outpacing other developments, Sun has introduced a JRE plug-in for web browsers, which allows users to run Java applications from their browsers or directly from their computers. As a result both Netscape and Microsoft have dropped JRE from their latest browsers.

Versions of Java up to version 1.2 (i.e. version 1.0 to 1.1.8) are generally known as 'Java 1'. Later versions (i.e. version 1.2 to 1.4.1) are generally known as 'Java 2'.

3.9.2 What is Java 3D?

Java 3D is an extension to the Java programming language that creates a connection between the JRE and a computer's 3-D graphics support. Several versions of Java 3D have been released by Sun.

Java 3D applications have some similarities with VRML viewers (see Section 3.7). The main difference is that Java 3D applications are compiled programs. Unlike VRML, the source files which make up a Java 3D world (the three-dimensional objects, textures, sounds and interactions) are compiled together with the viewer into an application that users can run on their computers.

3.9.3 How to view Java 3D worlds

Users can run Java 3D applications either in their web-browser or directly on their computers if they:

- Install a compatible JRE plug-in; at present this means a JRE plug-in that is compatible with versions of Java that are more recent than version 1.2.1
- Install a Java 3D JRE plug-in that is compatible with the graphics capabilities of the computer
- Download a Java 3D application.

The process is complicated for users. Firstly, Java 3D applets will *not* work with Java-enabled browsers that do not have the version 1.2.1 JRE *plus* the Java 3D plug-in installed. Secondly, users must select and install versions of the JRE and the Java 3D plug-in that are compatible with each other. Finally, users are recommended to check the graphics capabilities of their computer before choosing between the Direct X or OpenGL versions of the Java 3D plug-in.

Java 3D applications can be run on Windows and Linux operating systems but cannot yet be run on Apple Macs, although it may become possible for the Mac OS X operating system in future.

3.9.4 What is needed to create Java 3D applications?

To create Java 3D applications developers need:

- A Java development environment
- A Java3D software developer's kit – which will incorporate both the JRE and the Java 3D plug-in.

Java and Java 3D can be written by hand using the Java and Java 3D Software Developer's Kits that are freely available from Sun. For an experienced computer programmer, this takes no financial investment other than the time taken to learn how to write the programming language and access to a computer. A number of tutorials are available on the World Wide Web from Sun and the Java 3D community web-sites. However, learning to develop robust applications could take a novice from six months to two years. Writing programs by hand is also time consuming and the process of finding errors and debugging the code can cause hours of frustration.

Another option is to use a development environment, such as Borland's JBuilder, to ease the process of writing and debugging the program code. The three-dimensional objects that will make up the world are often created using a graphics package (such as AutoCAD, 3D Studio Max or Lightwave 3D), the objects are then converted into Java 3D code using a *loader*.

Java 3D developers may start by creating a world using VRML as a development environment and then converting it into Java 3D using a VRML loader. However, Java 3D does not support all of the features of VRML. In future X3D and the Xj3D browser, being developed by the WEB3D Consortium with Sun, may offer an alternative method (see Section 3.8).

Java 3D applications are compiled programs, which means that Java 3D worlds are hard coded with the viewer. For developers this meant that Java 3D worlds could not be output to file, edited and relaunched. As a result, the development process involved working with the 3-D models and other files that comprise the world in their original file format. With adequate documentation, worlds could be reloaded into Java 3D. Recently, Sun released *Java 3D Fly Through helper classes* which will allow developers to output Java 3D models in j3f format files and should ease development. However, the best preservation strategy (see Section 6) is to maintain back-up copies of the original files with adequate documentation.

3.9.5 Why don't people like Java 3D?

There are several limitations to Java 3D. Java 3D applications typically run more slowly than some other VR viewers. Large VR models may not work, as Java implementations may limit the amount of memory that can be used. The Java 3D plug-in is not available for Apple Macs, so applications will not run on all computers.

Perhaps the main reason why users dislike Java 3D is that it is complicated to install. Two different plug-ins must be downloaded and, as the JRE and the Java 3D add-on are large files, this can take a long time for a user who is connected to the Internet via a 56k modem. Finally, having downloaded the files and installed the plug-ins, Java 3D does not offer the same interactivity as other VR software programs.

3.9.6 Why do people like Java 3D?

Java 3D allows developers to create VR applications on a relatively low budget that can be delivered across a number of platforms. The development tools are inexpensive or free. Developers can also make use of Sun's Java Web Start application to provide software updates to their users through the Internet.

These factors may make Java 3D more suitable for in-house work rather than web-delivery. In-house work allows the developers to control the delivery platform, but delivery over the web means that developers are dependent on the end-user's willingness to install plug-ins.

How to find out more

Most information about Java 3D is available from www.java.sun.com (http://www.java.sun.com/products/java-media/3d/) and at the Java 3D community (http://www.j3d.org/) site. The Java 3D community web-site includes lists of frequently asked questions (FAQs), tutorials and articles about Java 3D. Javaworld magazine (http://www.javaworld.com/) includes a number of articles on Java 3D.

Resources:

Sun do not currently make the source code available for Java 3D.
- JRE plug-in from www.java.sun (http://java.sun.com/j2se/1.3/jre/)
- Java 3D plug-in from www.java.sun (http://java.sun.com/products/java-media/3D/download.html)
- Java Software Developers Kit (http://java.sun.com/j2se/)
- Java 3D SDK (http://java.sun.com/products/java-media/3D/download.html)
- Borland JBuilder (http://www.borland.com/jbuilder/)
- File Loaders (http://www.j3d.org/utilities/loaders.html)
- Java 3D Fly Through (http://java.sun.com/products/java-media/3D/flythrough.html)

3.10 OTHER ENVIRONMENTS

There are several other virtual reality development environments. This is a rapidly changing market place and this section will look at some VR products which have recently been withdrawn because they have been superseded, failed to live up to their initial promise or were not commercially viable.

When selecting a product, VR developers are recommended to consider which formats are supported and which plug-ins are required to view their worlds. This should help to assess whether users will be able to continue to view their virtual worlds should the development environment be withdrawn.

3.10.1 Superscape

Superscape DO3D, 3DWebmaster and VRT were VR authoring software developed by Superscape (http://www.superscape.com/) for creating 3-D worlds. All three had a world editor with a point and click interface for construction of virtual reality worlds and a large library of shapes, objects, sounds and textures. DO3D was aimed at home use and was limited to creating worlds from the library using simple shapes. 3D Webmaster had a shape editor to allow construction of user-defined shapes which could then be used in the world editor. It was possible to create hotlinks easily and integrate the worlds into 3-D web-pages, a Java interface allowing two-way communication between the 3-D web-page and Java applets. Superscape Control Language (SCL) allowed behaviours and interaction to be programmed into worlds. VRT was the high-end product; it added the capability of building stand-alone applications for Windows (viewed using Superscape's Visualiser program), adding libraries of code written in C and had three extra editors for customising keyboard control, menus and dialog boxes and designing user interfaces.

All three authoring tools supported SVR, Superscape's 3D graphics format, and VRML97. Unfortunately all have now been withdrawn from the market by Superscape. The Viscape viewer (used to view models built using these tools) is similarly no longer available for download from Superscape.

3.10.2 Microsoft Chromeffects

Chromeffects was seen as Microsoft's answer to VRML. With Chromeffects, Microsoft aimed to integrate virtual reality into web-pages and to offer a virtual reality authoring tool as part of the Windows platform. A software developer's kit was released but, following feedback, Microsoft withdrew Chromeffects pending more development work. Software developers had asked for better compliance with World Wide Web Consortium standards.

Section 4: Collaborative Virtual Environments

This section is partly based on the article 'VRML Multi-User Environments' by Rachael Edgar and Ben Salem (Edgar and Salem 1998), researchers at the Networked Virtual Reality Centres for Art & Design.

4.1 CVE TECHNOLOGY: HOW DOES IT WORK?

The World Wide Web operates on the client-server model. This means that, for any communication to take place, one machine will adopt the role of a *server* and the other that of a *client*. Servers are set up to serve formatted files (using extensions such as 'html', 'jpg', or 'gif'), and clients are set up to receive them.

3-D on the web is served up in the same way as other file formats. The file extension for VRML files for example is 'wrl' and, once the server has transferred a VRML file, the browser software on the client side is activated. The browser builds, or renders an internal model of the 3-D world and displays an image of it on the screen. Whenever the user changes viewpoint within the world, the image is re-rendered accordingly.

Collaborative Virtual Environments (CVEs) are networked virtual reality systems, enabling groups of people to come together in virtual space. CVEs actively seek to support human-human communication and collaboration in addition to human-machine interaction.The participants are *virtually* in the same shared virtual world and can interact with it and with each other, represented by their virtual embodiment, an *avatar* (see Section 4.2).

The concept of actually *sharing* a three-dimensional world poses a number of problems that other, static, types of web media do not. Actions, reactions, geometry, images, positions, orientations, and conversation all have to be shared and passed across the network. The simple act of tracking the position and state of a number of avatars and passing this information out to a number of users can cause immense problems.

CVE systems typically try to minimise the information that is exchanged between server and clients only to that which is absolutely necessary. When entering a CVE there is an initial download of the bulk of the world description and then short bursts of communication as information is passed to and from the server.

Designing worlds for collaborative virtual environments is similar to creating single-user 3-D spaces. Typically, the server requires some additional identification information in the 3-D description file. In VRML files, this could be as short as a single line specifying the location of the server system, typed in with a normal text editor. In principle, every VRML world could be made multi-user accessible. Therefore, the same guidelines that apply to virtual reality methods and techniques also apply to CVE content creation.

4.2 USER REPRESENTATION

In collaborative virtual environments, users are generally represented by a virtual embodiment, commonly referred to as an avatar. An avatar represents a user's awareness and identity within a virtual space, allowing interactions to take place. In CVEs, an avatar allows a user's movements and whereabouts to be seen by other inhabitants of the virtual space.

Figure 2: Cartoon avatar from Blaxxun

In most on-line communities there is a choice of avatar, although this choice may be large or small depending on the software and the world visited. Avatars may be realistic representations of human beings or creatures that are more cartoon-like. They may be rigid objects that move around as if they are on rails, or animated virtual bodies that can move or gesture. Users may be able to design their own avatar and then use it in the community of choice. Most on-line CVE providers will support avatars designed in VRML (See Section 3.7).

An avatar provides a way for others to understand a user's intended persona and identity. Avatars also help users to establish a feeling of actually 'being there', of self-location or presence in the virtual world.

Figure 3: Customised avatar from ActiveWorlds

Figure 4: Humanoid avatar from Blaxxun

In systems that support communication between participants, an avatar can provide direct feedback about one particular user's actions, attention and current state to the other inhabitants. Avatars can be used to convey the user's emotions through gestures, postures or even facial expressions. For example, a user might indicate elation by making their avatar jump up-and-down. However, a jumping avatar could mean that another user is trying out some of the buttons on the browser's interaction panel!

4.3 INTERACTION AND COMMUNICATION

The avatar (see Section 4.2) is a user's main device for interaction within the 3-D environment. Many of the avatars in on-line communities have predefined scripts which, when executed, cause the avatar to move or gesture in a certain way. The gestures available are most commonly such things as waving, dancing, smiling or, in other words, actions or emotions that are easily conveyed by movement.

Gesturing is not the only way to communicate. All on-line communities have chat-boxes and chat-logs. A chat-log shows the conversation between all users and a chat-box is where a user enters a contribution to the conversation. Some chat tools have features that give users control over the conversations that they take part in. For example, messages from all users in the same room (area or world) are usually displayed for all to see. But some chat tools allow users to prevent messages from specific individuals from being displayed in their chat window. Other chat tools allow private conversations between users and offer a 'private chat' mode where the conversation is invisible to the other inhabitants.

Users are easily distracted from the 3-D environment when writing and reading chat messages. In general, once inhabitants have started to be engaged in a conversation, there isn't much movement or interaction with the world. This is a paradox for developers, who hope that users will interact using both the 3-D interface and the chat tools.

The Virtual Reality Case Study Library includes some guidelines on the 'netiquette' of collaborative virtual environments.

4.4 SYSTEMS OVERVIEW

Two Collaborative Virtual Environments are compared for this guide, ActiveWorlds and Blaxxun. These are both on-line communities with a more or less commercial interest, accessible via the World Wide Web. Another system, not compared in this Guide, is Deepmatrix (http://www.geometrek.com/products/deepmatrix.html) from Geometrek. Deepmatrix is an open-source system that enables anyone with some Java and VRML skills to design and maintain avatars and shared worlds on their own on-line community server.

Two projects using CVE technology are included in the Virtual Reality Case Study Library:

- CyberAxis, a three-dimensional virtual gallery with multi-user access. CyberAxis is based on the Axis database, the national register of contemporary British artists, and offers a completely new interface to browse the database
- Building Babel II, an experimental workshop which took place over three days in September 1998 at Coventry School of Art and Design. The workshop was an exploration of the issues surrounding the construction and use of CVEs, specifically by the art and design community.

In addition, there are systems that exist as prototypes in the academic environment, usually requiring high-end graphics computers for their development. Such prototypes have influenced the development of the on-line communities and two in particular should be mentioned:

- The DIVE system (http://www.sics.se/dive) is an experimental platform for the development of multi-user environments. Launched by The Swedish Institute of Computer Science (SICS), it focuses on interaction with and the behaviour of objects in a virtual world
- MASSIVE (http://www.crg.cs.nott.ac.uk/research/systems/MASSIVE/) is a distributed virtual reality tele-conferencing system developed by the Communications Research Group at Nottingham University. MASSIVE allows groups of people to hold meetings in a virtual conference room, supported by real-time audio communication and shared workspaces in the form of blackboards.

4.5 ACTIVE WORLDS

4.5.1 What is Active Worlds?

Active Worlds is a company which offers software products and on-line services. The company hosts collaborative virtual environments that have been developed using its software. For example, AlphaWorld is a virtual 'real estate' in which users can create and visit virtual structures. Free software is available from Active Worlds (http://www.activeworlds.com/) that is straightforward to install and clear instructions are available. 'Active Worlds' is a browser in its own right and does not rely on either Netscape Navigator or Internet Explorer. AlphaWorld provides users with access to more than 1000 inhabited worlds, many of which are commercial and incorporate shopping facilities.

4.5.2 Using Active Worlds

Users can participate in 'Active Worlds' on two levels: tourist and citizen. Tourists can visit and chat without charge at anytime. Citizens pay an annual fee in order to be able to send and receive 'telegrams' and to be able to 'build' on a plot of land. Tourists have a choice of a male or female avatar; citizens have a greater choice of avatars offering a wider range of expressions.

A chat window is located below the 3-D browser window and in addition messages are displayed above the heads of each avatar. This feature means that users don't have to check the chat window all the time. The physical means of expression depend on the avatar, but can include ANGRY, HAPPY, POINT, SHOVE, DANCE, WAVE, or FIGHT. Active Worlds offers quite sophisticated animations but, unlike Blaxxun (see Section 4.6) where users can design their personal avatar, in Active Worlds the avatars and their abilities are pre-defined.

An interesting feature of Active Worlds is the web page window that appears at the side, giving information about the world and its community. Citizens can create links from a plot of land to pages, images, and other multimedia files that appear either in the 3-D space or in the browser window at the side.

4.5.3 Creating a world

Citizens can claim their own piece of cyberspace and build a 'home' on it which can be visited by other users. Active Worlds incorporates a library of pre-defined elements which provide the building blocks with which citizens can create a house, castle, spaceship etc. The company also works with commercial clients to develop worlds for them. Active Worlds has launched an 'Education Universe' to make their technology available to educators at lower costs.

For more detailed information, take the ActiveWorlds Guided Tour (http://www. activeworlds.com/tour/index.html) or visit the ActiveWorlds University (http://www. awcommunity.org/awu/) which offers citizens free classes in building worlds.

4.6 BLAXXUN

4.6.1 What is Blaxxun?

The Blaxxun Platform (http://www.blaxxun.com/) is a modular software system that can be used to create virtual worlds based on VRML97. It is a browser which can be accessed via HTML, Java or a plug-in and runs on Linux, SGI, Sun and Windows NT platforms. The Blaxxun Platform offers users communication via instant messaging, message boards, calendars, surveys, voting and SMS alerts. The system allows collaboration with shared applications incorporating audio, video and text-to-speech with 3-D visualisations of products and environments. Blaxxun also provides facilities for those who create and administer worlds, for example the management of user profiles, roles, access rights and user accounts. Worlds can incorporate features to attract users such as clubs, meeting places and personal home pages.

4.6.2 Installation

Users need to accept the licence agreement and install the *BlaxxunContact* plug-in and can choose between an automatic or a manual installation. The automatic installation routine checks the user's hardware and software and selects the appropriate version of BlaaxunContact.

Visitors navigate in worlds using their mouse or keyboard or can take a quick tour (if one has been provided by the developer). They have the option of selecting an avatar from a library and, in some worlds, can use an avatar that they designed themselves. Visitors can contact each other by selecting a person from the people panel and 'beaming' to them. They can then chat or choose to ignore each other.

4.6.3 Developing applications

Blaxxun provides a full application programming interface (commercial users require a special licence). Developers can create a Blaxxun community using the API by first designing a world in VRML 97 and then using a template to set up a web-page and link this to a series of files which provide the multi-user functionality. The world is then established as a multi-user meeting point. Developers can customise this meeting point by designing their own avatars and automated agents to greet visitors or by creating shared events advertised to all visitors.

Developers of more complex applications may choose to install their own Blaxxun community server.

Blaxxun Contact supports the full VRML 97 specification and also incorporates Java code. The Blaxxun source code is available through the Web3D consortium.

4.6.4 Finding out more
- BlaxxunContact download (http://www.blaxxun.com/services/support/download/install.shtml)
- Blaaxun Developer Guide (http://www.blaxxun.com/support/developerguide/developer/index.html)
- Web3D Blaxxun Source Code (http://www.web3d.org/TaskGroups/source/blaxindex.html)

- Case Study 7: CyberAxis
- Case Study 8: Building Babel II

Section 5: Documenting Data from a Virtual Reality Project

Documentation is important. When subject specialists think about documentation they tend to think of the final report presenting the history, techniques and results of a project. But virtual reality projects involve a number of people, the capture of data from various media and the creation of worlds for an audience. Although the report is an important part of the project archive, additional information about factors which influenced the development, or that support the maintenance and future reuse of the world, should also be documented.

It is important to prepare technical documentation about the world. This documentation should describe the steps taken in creating the world and record which file formats are in the archive, the software applications used, the intended user community, the software used to view the world, and so on. For worlds that are based on reconstructions it is important to record which elements are based on real evidence (real-world measurements of monuments and buildings) and which elements are artistic interpretation (Miller and Richards 1995; Eiteljorg 2000).

When combined with the report, such documentation enables the effective description of the resource. Documentation is an essential prerequisite for depositing project data in a digital archive.

5.1 PLANNING FOR THE CREATION OF DIGITAL DATA

From the moment a project begins, careful thought must go into the digital files that will be produced and their life-span. Planning should include:

- Preparing a project design that documents the tasks necessary for the successful completion of the project at its outset and includes a summary of the types of digital data that will be created and the projected lifespan of the virtual reality. It is important to update this documentation throughout the life of the project.
- Defining and documenting the individuals, groups or organisations that are responsible for creating and managing digital files at all stages of the process.
- Planning which file formats will be used for both the dissemination of data and for archiving purposes. Different formats may be used for these two activities. At the planning stage of the project, it is important to check the guidelines or standards recommended by the digital archive facility destined to receive the files and ensure that these are followed.

Project managers are recommended to consult an archive facility such as the AHDS for up-to-date information when considering the archiving of virtual worlds.

5.2 PROJECT DOCUMENTATION

This section looks at documentation for the project itself and is based on Dublin Core metadata (see Miller and Greenstein 1997). The documentation recommended provides a record of the project team, the background to the project and the archive that has been created. It is important because it supports discovery (see Section 7.1 and Section 7.2) of the resource on the World Wide Web and in the digital archive.

The types of information recommended for inclusion in the project documentation are listed below.

Title
The name of the subject, project or model. It may be either the name used for the published model (e.g. 'Dancing in Virtual Spaces') or the familiar and/or published place or monument name (e.g. 'Wroxeter').

Creator
The name and address of the organisation or individual(s) who created the world. Include the name(s) and address(es) of those who have made a significant contribution to the world, for example educationalists, artists, translators.

Subject
Keywords or phrases that describe the subject or content of the resource. This may include:
- Subject discipline; e.g. Archaeology, Architecture, Fine Arts, Visual Arts, Theatre Studies
- Subject type; Fort, Dance, Sculpture etc.
- Temporal period; e.g. Roman, Medieval

It is recommended practice to make reference to controlled vocabularies or classification schemes when indexing the subject content of the model (see Appendix 3).

Description
A brief summary (max. 300–400 words) of the main aims and objectives of the project for which the model was developed and a brief description of the model and its features. The projected lifespan of the virtual reality should be noted.

Publisher
The name(s) and address(es) of the organisation or individual(s) who have facilitated access to the resource. This may include:
- The organisation or agency who funded or grant-aided the creation of the resource
- The client who commissioned the creation of the resource
- The person or organisation who deposited the resource in an archive repository.

Date
Dates associated with the creation and dissemination of the resource. Useful dates include:
- Project start and end date
- Release date for the electronic resource
- Dates associated with the lifespan of the resource, e.g. maintenance cycle, update schedule.

Type
This is the general form of the resource, e.g. virtual reality presentation, electronic learning resource, digital archive.

Format   The data formats which make up the resource, e.g. html, vrml, jpeg image, mpeg.
Identifier   The identification number or reference used to identify the resource. This
                    may be an internal project reference number.
Source   References to the original material for any data derived in whole or in part
                    from published or unpublished sources, whether printed or machine-readable.
                    Related archives may include plans or section drawings which have been
                    used to create 3-D graphics.
                       Details should be given of where the sources are held and how they are
                    identified there (e.g. accession number). If a digital collection is derived from
                    other sources there should be an indication of whether the data represent a
                    complete or partial transcription/copy, and the methodology used for its
                    computerisation. Also full references to any publications about or based upon
                    the modelling project should be provided.
Language   The language(s) of the intellectual content of the resource.
Relation   Record the relationship to other resources, for example:
- Bibliographic references to publications about the project
- The name and address of the organisation or individual(s) holding the primary data for the world. Primary data may include image files, 3-D graphics files, sound files etc.

Coverage   Where the model relates to a real world location it is useful to record
                    information about its spatial coverage. Where appropriate this may include:
- The country in which the model lies
- The current and contemporary name(s) of the country, region, county, town or village
- The administrative areas (e.g. County, District, Parish) to which the world relates
- The map co-ordinates of the SW and NE corner of a bounding box enclosing the study area. For Britain, Ordnance Survey National Grid co-ordinates are recommended.

Use of the standard area names from an appropriate documentation standard is recommended (see Appendix 3).

Rights   A description of any known copyrights held on the model or the source
                    materials used in its creation.

5.3 DOCUMENTING THE AUDIENCE

In addition to project documentation (see Section 5.2), it is important to record information about the intended audience for the virtual reality resource. This documentation will help to guide the development of the resource and also enables later resource discovery and reuse. Such information is particularly important for virtual reality applications which are developed for education or training purposes. The Dublin Core Education working group (http:// dublincore.org/groups/education/) proposes that the following information should be recorded:

Audience   A description of the type of user for which the resource is intended. For

	educational and training resources state the category/academic level of the intended users, e.g. sixth-form students.
Mediator	For educational or training resources that are presented to an audience by an individual or organisation, describe who mediates a user's access to the resource, e.g. museum interpreter.
Education standard	For resources that have been developed to conform to an established academic or process standard, a url or publication reference for that standard should be recorded, e.g. English National Curriculum Key Stage 4, http://www.nc.uk.net/
Interactivity type	Describe the type of interactivity that the end-user has with the resource.
Interactivity level	Describe the level of interactivity that the end-user has with the resource.
Typical learning time	Describe the typical length of time that it is anticipated that an end-user will spend interacting with a resource.

5.4 DOCUMENTING METHODS AND TECHNIQUES

It is important to record information about the techniques used during the virtual reality project. This will help with both maintenance and testing of the model and in archiving it for subsequent reuse.

5.4.1 Documenting the virtual reality application

The following is a list of information that might be useful to record:

Model type	The form of the virtual reality should be recorded. The following list, while not exhaustive, gives examples:
	• panorama
	• bubble world
	• vrml model
	• collaborative virtual environment.
Application format	Specific information about the VR application is required and the version used should be recorded, e.g. Java 1.3.1 with Java3D 1.2.1.
Application specification	Reference should be made to a copy of the specification for the correct version of the VR application that has been used. This may either be a reference to a published document (e.g. the URL for the specification on the World Wide Web) or a copy of the specification itself retained as part of the project archive. For ISO standards, it is sufficient to record the ISO document number and a URL, e.g. the specification for VRML 97 is ISO/IEC 14772-1:1997 and can be found at http://www.VRML.org/technicalinfo/ specifications/vrml97/index.htm.
Hardware platform	Record the specification for the computer system that has been used in developing the VR.
Authoring tools	Any software that has been used in creating the world, e.g. AC3D, Borland JBuilder. Note the version number of the authoring software and the VR format that it produces, e.g. VRML 97.

3-D drawing tools	Record any 3-D drawing packages (e.g. AutoCad, 3-D studio etc.) that have been used to create objects for incorporation into the world. Note the version number used, the file format produced and the method used to incorporate objects into the world.
3-D scanners	If a 3-D scanner has been used to capture the geometry of an object, record the make and model of the scanner that has been used and the resolution at which the object has been scanned.
Object libraries	If objects have been obtained from object libraries, it is important to record: • the object(s) • the object library or source • the copyright of the object(s) from that source.
Animations/ Scripts	Record the scripting language that has been used, e.g. Javascript, Java 1.2
Sounds	It is useful to describe how sound has been used in the world, how the sounds have been obtained, the file format used (e.g. MIDI, WAV) and whether they have been compressed for incorporation into the world.

5.4.2 Additional documentation for images

Images may be created for use in virtual reality applications for backgrounds, panoramas, surface textures and so on. They may be created using computer-aided drawing packages or captured using digital cameras or 2-D scanners. Original images may be created using high-quality data capture techniques and then compressed for dissemination in the virtual reality application. In some cases it may be advisable to archive multiple versions of each digital image with appropriate documentation.

Images/ Textures	Record the methods used to create surface textures, for example if a texture is based on a photograph of the real world object, a similar real world object, or if the texture has been created by a designer for the purpose and so on. It is useful to record not only the file format(s) in which the original images were created or acquired but also the format in which they have been incorporated into the world after processing or compression (e.g. BMP, GIF, JPEG, TIF).

5.4.3 Additional documentation for reconstructions

Virtual reality worlds may be created partially from survey data captured from real-world objects (e.g. using a 3-D scanner) and partially through interpretation and artistic licence. It is important to document which portions of the model are which. This avoids misleading users and can help to enable the original survey data to be reused in alternative interpretations.

Real-world objects	If objects have been created from real-world evidence, it is important to record: • object(s) • the source file(s) • the person or organisation(s) responsible for the survey

- the techniques or equipment used, e.g. 3-D object scanner, electronic distance measure, topographic survey, hand measurement etc.
- the date(s) when the survey took place
- a brief description of the survey and the area covered
- any associated rights
- any conventions used to depict real-world objects in the VR.

See CAD: a Guide to Good Practice for more information about capturing data for CAD models from field survey.

Interpretative objects   If objects have been created to present a particular reconstruction or interpretation, it is important to record:

- object(s)
- the source files
- the person or organisation(s) responsible for the interpretation
- a brief description of the interpretation and the evidence on which it is based
- references to any bibliographic sources which support the particular reconstruction
- the date of the reconstruction
- any associated rights
- any conventions used to depict the interpretative objects in the VR.

5.5 DOCUMENTING THE DELIVERY PLATFORM

It is useful to document the delivery platforms on which the world has been tested. This information helps with both maintenance and testing of the model and in archiving it for subsequent reuse.

It is useful to document the following information, for every delivery platform that has been tested.

Operating system   Record the operating system on which this delivery platform runs, e.g. Windows 2000.

Browser software   Record the browser version(s) on which the world has been tested, e.g. Netscape Navigator 4.75.

Plug-in/viewer   Record any plug-in(s) or viewer(s) required to view the world on this platform. Include details of the name, address and URL of the source from which the plug-in or viewer can be obtained. It is useful to describe how the plug-in should be installed and note any optional settings.

Scripting languages   Document the scripting language and version number used in the world and describe how it is implemented in the delivery platform.

Hardware system   It is useful to document the minimum hardware specification required to run the world, i.e. the processor, RAM and graphics card.

Network connection   It is useful to document the minimum network connection required to run the world, e.g. a dial-up 56 kbits modem, 2 mbpits broadband connection.

5.6 REPORTING PROJECT OUTCOMES

Each virtual reality project should result in a report that describes the project and its outcomes. The report is an important part of the project documentation as it provides a description of how the virtual reality was produced and its look and feel. It also offers an opportunity to record the audience response to the virtual reality – their engagement and the successes and failures of the techniques that have been used.

For some virtual reality installations, for example exhibitions taking place in collaborative virtual environments, the report provides a record of a performance or an event.

Reports should contain:

- an abstract or summary
- an introduction to the project and its objectives
- description of the methods
- description of the look and feel of the VR produced
- the projected lifespan of the VR
- the results
- a record of any key events
- conclusion.

Reports should also contain an appendix which provides an index to:

- the media files which make up the VR and their location
- details of how the media files relate to each other.

5.6.1 Documenting the report

The report itself and some basic cataloguing information should be deposited as part of the archive for virtual reality projects. The information required to document the report is:

Report title and reference number	The title of the initial paper or digital report that has been generated from the results of the survey, and any surveyor's reference number.
Report author	The name of the author(s) of the report.
Report holder	The name of the organisation or individuals who can provide copies of the report.
Report summary	A brief summary of the project; in most cases this will be the abstract from the report itself.

5.7 DESCRIPTION OF ARCHIVE

Without documentation, only the developers of a virtual world are confident of its structure – which files belong together, what their names mean and what software was used for their creation. If this information is not carefully documented it will be difficult to preserve the resource for the future. For example, if no information is available on the software used to render the virtual world, it will be virtually impossible to archive it for the future (see Section 6.2).

To request detailed information on each file is impractical; however, a simple list of all files with more detailed descriptions of related files should be possible. The following information should be provided with a digital archive:

List of all file names A list of all digital files in the archive, with their names and file extensions (e.g. 'NWPalaceTR.WRL', 'stone.mov').

Explanation of codes used in file names A brief explanation of any naming conventions or abbreviations used to label the files.

Description of file formats An explanation of which internal format is associated with a particular file extension (e.g. '.WRL files are VRML '97 files').

List of codes used in files A list of any special values used in the data (e.g. '999 indicates a "dummy" value in the data').

Date of last modification The date of last data modification allows the currency of the archive to be assessed.

5.7.1 File-naming conventions

Digital files should be given meaningful titles that reflect their content. It is recommended that standard file-naming conventions and directory structures should be used from the beginning of a project. If possible, the same conventions should be used for all projects by the same organisation, for example:

* Reserve the 3-letter file extension for application-specific codes, e.g. WRL, MOV, TIF
* Identify the activity or project in the file name, e.g. use a unique reference number, project number or project name
* Include the version number in the file name where necessary.

5.7.2 Version control

It is extremely important to maintain strict version control when working with files, especially with data which may be saved and processed using a series of different treatments.

There are three common strategies for providing version control: file-naming conventions, standard headers listing creation dates and version numbers, or file logs. It is important to record, where practical, every change to a file no matter how small the change. Versions that are no longer needed should be weeded out, after making sure that adequate back-up files have been created (see Section 6.2).

Section 6: Archiving Virtual Reality Projects

This section of the Guide explores the possibilities for archiving virtual reality for use beyond the immediate lifespan of a project. It considers the reasons that an individual or project team might have for considering archiving and discusses issues relating to access and reuse of virtual reality resources.

6.1 INTRODUCTION TO ARCHIVING VIRTUAL REALITY PROJECTS

Six months is a long time in the world of computing. Twenty years ago few people had seen a computer, except perhaps a BBC microcomputer or Sinclair ZX. Computers looked and behaved differently. We now work with 'user interfaces' that offer us visual ways of interacting with computers, for example the Windows Operating System. The computer hardware and programs that we use have also changed dramatically. There are three different layers to the way that we use computers: the hardware, the operating system and the software. These three layers could all change radically in the next ten years.

As Information Technology continues to change and grow, it is becoming increasingly important for projects to archive their work in order that a permanent record remains for the future. The world of computing tends to look to the future, rarely stopping to consider the present let alone the past.

This is especially true in the experimental world of Virtual Reality and related technologies. Projects are often seen as transitory experiments and little thought is given to reuse or to their place in the history of VR. One only has to read much of the literature about VR and its origins to realise that projects have physically (in terms of equipment) and digitally (in terms of files) disappeared or been broken up. This is true both of projects that were important in the history of the technology and of projects of importance to the discipline for which they were created.

6.1.1 Practical issues

Perhaps the best way of preserving virtual reality for the future is to consider archiving material from the start of a project and not just at its completion. Archiving virtual reality concerns not only the files that make up the world but also the original data files and supporting documentation such as the project report. The best strategy is for all of these digital data to be systematically collected, maintained and made accessible to users operating in very different computing environments.

Digital archiving is different from traditional archiving practice which seeks to preserve physical objects (e.g. artefacts, samples, paper, photographs, microfilm) that carry information.

Digital archiving is about preserving information regardless of the media on which that information is stored. This is because disks and other magnetic and optical media degrade and software and hardware change rapidly. Digital data are transferred from one storage media to another and they may be viewed using new generations of hardware and software.

6.2 DIGITAL ARCHIVING STRATEGIES

Back-up is the familiar task of ensuring that there is an emergency copy (or a snapshot) of all data held separately in case of damage to the original data (by accident or through a disaster). For a small project this may mean a single file held on a floppy disk or on a network; for a larger project or dataset it may involve complex procedures involving disaster planning, with fireproof cupboards, off-site copies and daily, weekly and monthly refreshing. Such back-up strategies are important in the lifespan of the project but are not the same as long-term archiving of the data.

Digital archiving does not rely on the preservation of a single disk, tape, or CD-ROM. The essence of digital archiving lies in one of three strategies (Beagrie and Greenstein 1998):

- Migration of information from older hardware and software systems to newer systems
- Emulation of older hardware/software systems in newer systems. This is technically challenging and becomes increasingly difficult as current technology becomes ever more remote from the original systems employed
- Complete preservation of old hardware and software systems. This very high-risk strategy should only be considered as a short-term measure as it depends on retaining the skills and resources needed to maintain and run the original system.

Data migration is the strategy recommended for most applications. Where standard data formats are available (such as ASCII text, TIFF images or VRML) data migration is successful in preserving data for future use. Where data cannot be migrated or the original 'look and feel' is of *substantial importance* a strategy of either emulation or technology preservation may be justifiable.

The Guggenheim's Variable Media Initiative (http://www.guggenheim.org/variablemedia/) is of particular interest to artists working with virtual reality. This initiative is exploring preservation strategies for ephemeral media and for the Guggenheim's collections of Conceptual, Minimalist and video art. Artists, museums and media consultants are working together to consider the implications of 'storage', 'emulation', 'migration' and 'reinterpretation' strategies for artworks. The Case Studies include the 1991 interactive installation of black rod liquorice candy by Felix Gonzalez-Torres. Although not virtual reality, this is of direct relevance as it explores the significance of different preservation strategies and the importance of lifespan, physical appearance and meaning to the work.

6.2.1 Practical issues

No digital archivist can successfully preserve data that are not fully documented; with every strategy there is the potential for information to be lost. Detailed documentation (see Section 5) allows archiving strategies to be carefully planned and tested in advance. Digital data also need to be regularly managed. Archivists use data management tools (such as Electronic Document Management Systems) to inform them when files held in deep storage facilities require active intervention. These are usually databases that, ideally, flag dates and automatically inform the system manager when files need attention.

6.2.2 Data migration

Digital archiving generally revolves around a policy of controlled data migration. Archiving by this strategy involves four main activities: data refreshment, data migration, documentation and data management tools.

Data refreshment is the act of copying information from one medium to the next as the original medium nears the end of its reliable lifespan. Research into the lifespan of both magnetic and optical media has shown that the former can be safe for 5–10 years and the latter may survive more than 30. However, technology changes much more quickly and digital media are far more likely to become unreadable as a result of changes in hardware and software than through media degradation.

The process of data refreshment involves copying data to new media as technology evolves. For example, data collected on 3-inch Amstrad diskettes might have been refreshed on to 5.25-inch disks and again on 3.5-inch disks as computer hardware developed.

Computer software changes even more rapidly than computer hardware. Data files that have been created in the proprietary format of a particular software package may not be retrievable in future. The software company may change the formats used in subsequent versions of the package, or may cease trading and the file may not be accessible by software produced by other companies. **Data migration** is the act of copying data files from one format or structure into another. For example, copying a word-processor file into a newer format while maintaining its original content and appearance. In some cases, migration may offer improvements in access to information, for example migrating an old CAD file to newer versions might allow users to access the enhanced functionality of next-generation CAD software.

File formats that have been identified as *international or open standards* support the migration of data files into new generations of software. This is because the definitions for these formats are published (e.g. VRML'97 is published by ISO (http://www.iso.ch/iso/en/ISOOnline.frontpage)) and they are implemented consistently by different software manufacturers. File formats that have become *industry standards*, because they are widely used, generally allow files to be imported into other software (e.g. DXF is developed by AutoDesk). Industry standard file formats can be interpreted differently by the various software manufacturers, which can mean that a file produced in one package cannot be read in another. However, industry standard formats generally can be migrated if archival guidelines to use the basic export are followed (see Section 6.4).

Some software manufacturers develop *proprietary file formats* for use in their own packages. In some cases, the definitions for the formats are available for other manufacturers to use. But in other cases, formats are used by a single software manufacturer and no definitions are available for others to use (e.g. Superscape developed the SVR format and Sun have developed JXF). Such files can generally be migrated into the next generation of the software package in which they were developed. But data held in proprietary file formats is vulnerable and may be lost if the parent software package ceases to be available.

Successful data migration relies upon good documentation and careful planning, testing and execution to preserve the original data (Wheatley 2001). Data management tools help digital archivists to plan for the tasks involved while good documentation helps them to understand the structure of the data fully and how the different parts relate to one another. Of equal importance is the use of open or industry standard file formats which allow migration to be executed without data being lost.

6.2.3 Emulation

Emulation revolves around using current technology to mimic the original environment. The digital project data are archived together with the original software and operating system. Future use of the data may involve mimicking either the original software or hardware on current equipment. The aim of emulation is to re-create the look and feel of the original system. Emulation can be technically challenging and may become more difficult over time.

The use of emulation for digital preservation is a very new area and, although it may not be a practical strategy at the current time, it may be used more widely in the future. Successful emulation will rely upon detailed documentation of the system as a whole and careful planning.

6.2.4 Technology preservation

Technology preservation is akin to traditional museums and archive practice. This strategy revolves around preserving the original hardware and software on which the system was installed. Future use of the data depends on maintaining and running the original system. This becomes increasingly difficult with time as the skills needed to maintain or repair the system are lost. For this reason, technology preservation is a very high-risk strategy and should only be considered as a short-term measure for digital archives.

6.2.5 Virtual Reality

All three digital archiving strategies have relevance when archiving virtual reality. A migration strategy can be used for virtual reality developed using standard formats, for original data files, screen-shots and associated documentation. Where it is important to preserve the 'look and feel' of the original virtual reality, emulation may be appropriate. For virtual reality applications that are dependent on specific hardware and software, emulation may be the only option. Technology preservation is not sustainable in the long term.

Archiving virtual reality is developing with the technology. It presents technical challenges and it is difficult to predict future archival strategies. It is important to consider the lifespan of each project. Some projects involve research into new techniques and the virtual reality that is developed may have a limited lifespan. Other projects have a longer life-span. If researchers wish to preserve virtual reality for the future, the best strategy is to adopt standard formats.

6.3 ARCHIVING PARTICULAR VIRTUAL REALITY FORMATS

As virtual reality worlds are multi-media applications and standards are still evolving, archiving is more complex than with other forms of digital data. VR developers will need to consider carefully which aspects of a world should be preserved in the long term and the best way to go about this. If the *look and feel* of a virtual world is important, does preservation mean attempting to keep the world running on its original hardware and software (with the risks inherent in this strategy, see Section 6.2)? Or is it sufficient to preserve the original source files with screen shots of the world and detailed documentation of how the world was created?

The approach that is taken to archiving VR will depend on both the technology that has been used and the nature of the project. Where standard VR formats have been used, such as VRML and X3D, worlds can be migrated as computer hardware and software evolves. For projects such as archaeological reconstructions, archiving the original source files (the original images, CAD models and so on) used to create the world will be important. For many projects an acceptable alternative to archiving the VR world itself, which may be both difficult and uncertain for proprietary or non-standard formats, might be to break down the VR into its original source files. With this approach, the source files would be deposited in standard formats together with screen-shots of the world and a detailed description of how to put the elements back together to recreate the application. In other cases, archiving the *look and feel* of the world may involve emulation of the VR on future platforms.

Archiving virtual reality in any format involves depositing the data files that make up the application with the project documentation (see Section 5) and metadata records (see Section 7.2). Particular VR formats require the additional data and documentation listed below to allow for either emulation or re-creation of the application.

6.3.1 Bubble worlds and QTVR

For panoramas and bubble worlds, it is recommended that the following should be deposited as part of the project archive:

- Original image files with documentation of any 'hot-spots'
- The specification, version number and date of the format used. Use of industry standard formats (e.g. QTVR) is recommended
- A list identifying the name and version number of plug-ins or viewers which can be used to view the application
- A copy of at least one plug-in/viewer that can be used to view the application
- Installation instructions.

6.3.2 VRML

For virtual reality applications developed in VRML, it is recommended that the following should be deposited as part of the project archive:

- The VRML specification, version number and date should be identified. For versions other than the ISO standard, VRML 97, a copy of the specification is recommended
- A list identifying the name and version number of all VRML plug-ins or viewers which have been tested for use with the application

- A copy of at least one plug-in/viewer that can be used to view the application
- Installation instructions.

6.3.3 Java3D

Virtual reality applications developed in Java3D are not suitable for archiving by a data migration strategy except when archived as the original files. It is recommended that the following should be deposited with the project archive:

- The original 3-D models in archival formats
- Documentation describing how the world was constructed
- A report which illustrates the world and describes how it was used
- A copy of the correct version of the Java Developer's Kit (JDK) with the appropriate Java3D plug-in
- A copy of the appropriate versions of the Java and Java3D specifications
- A copy of the correct version of the Java Run-time Environment (JRE) and the appropriate Java3D plug-in
- Installation instructions.

6.3.4 Collaborative Virtual Environments

Collaborative Virtual Environments may not be suitable for archiving by a migration strategy except when archived as original data files. The following should be provided with the project archive for the purposes of migration:

- The original 3-D models (including worlds and avatars) in archival formats
- Documentation describing how the models were incorporated into the CVE and the interactivity
- Where appropriate, the 'chat-log' for an exhibition or event
- A report illustrating the CVE and describing how it was used
- Both client-side and server-side application software with specifications, documentation and installation instructions.

6.4 GUIDELINES FOR DEPOSITING DIGITAL ARCHIVES

In some cases, project funding carries with it an obligation to deposit the application that is produced in an appropriate digital archive. In other cases the decision to deposit a virtual reality project will be made by the project team in discussion with an archive repository.

There are several factors to be considered when proposing a virtual reality project for deposit in a digital archive. These include:

The importance of an application for a particular discipline

The importance of the materials being proposed for archiving will generally be considered in a peer review process. This enables both the depositors and the archive managers to clarify the intellectual problems concerned with the application in the context of the subject discipline.

This process should also assist in assessing the potential levels of interest in reusing the application in future.

The cost of archiving

The costs of archiving the application will vary according to how (or even whether) it can viably be managed, preserved and distributed to potential secondary users. Project managers are recommended to contact an appropriate digital archive at an early stage in the project to discuss their guidelines, not least because different archives have different methods of offsetting the costs of archiving.

Formats and documentation

Each archive has terms and conditions that the project must adhere to for depositing and archiving its material. Most archives will have guidelines covering the data formats that are accepted, the documentation that is required and terms and conditions for access. It is a good idea to discuss these at an early stage and to follow the guidelines from the beginning of the project.

6.5 DEPOSITING VIRTUAL REALITY WITH THE ARTS & HUMANITIES DATA SERVICE

The Arts and Humanities Data Service (AHDS) archives, disseminates and catalogues high-quality digital resources of long-term interest to the Arts and Humanities disciplines. Its geographical remit is to provide digital archiving facilities for all areas of the world in which UK academics have research interests. The scope of the AHDS collections is thus international.

The AHDS acknowledges the considerable benefit to both depositors and users of an effective and rigorous process of peer review of materials proposed for accessioning. In order to assist the AHDS to evaluate datasets and maintain the rigorous standards necessary for the effective development of a quality resource base, *Collections Evaluation Working Parties* have been set up by each of the AHDS service providers.

Data resources that are offered for deposit to the AHDS will be evaluated to:

- Assess their intellectual content and the level of potential interest in their reuse
- Evaluate how (even whether) they may viably be managed, preserved, and distributed to potential secondary users
- Determine the presence or absence of another suitable archival home.

Whereas the first form of evaluation involves assessment of the content of a data resource, the second focuses more on data structure and format, and on the nature and completeness of any documentation supplied. The third evaluation criterion is intended to prevent duplication of digital archiving efforts and to preserve the integrity of existing digital archives. Such evaluation is essential to determine how best to manage a digital resource for the purpose of preservation and secondary reuse, and also to determine what costs may be involved in accessioning and migrating the digital resource.

6.5.1 Deposit formats

In the last few years many data formats have appeared that are intended to make data exchange and migration easier. Some of these formats are proprietary (i.e. they are marketed by a single company) but many are open standards that are independent of the software that is used. In general, open data formats are preferred for digital preservation. Unfortunately, software manufacturers use open file formats but occasionally change them slightly so as to be less than 100% compatible with other software manufacturers. It is best to enquire with a digital archiving body in the planning stages of any project to check if there are any concerns with the anticipated file formats.

The format in which data will be deposited depends on the type of information that it contains. The file formats outlined in Table 2 are recommended for delivery, long-term preservation and for Internet dissemination. 'Delivery' formats pertain to the file types that will be accepted by the AHDS as a component of a deposit. Where necessary, these delivery file formats will be migrated into a 'Preservation' format for long-term storage and may also be converted to a 'Dissemination' format for delivery over the Internet. Dissemination formats may also include widely used proprietary formats such as Microsoft Word and Adobe Acrobat files for texts and jpeg for images, which may have no long-term preservation potential.

Data type	Delivery formats	Preservation formats	Dissemination formats
Virtual Reality	QTVR, VRML, X3D	VRML, X3D	QTVR, VRML
CAD	DXF, DWG plus some native file formats	DXF, DWG	DXF, DWG, DWF
GIS	ArcInfo, ArcView, DXF, DWG, MIF/MID, NTFF, SDTF, MOSS, VDF	ArcInfo, DXF, DWG	ArcInfo, DXF, DWG
Images	Uncompressed TIFF, GeoTIFF, BMP, PNG, JPG, GIFF, PCX, SVG	Uncompressed TIFF, SVG	JPG, GIF, PNG
Moving images	MPEG, Quicktime, Real Video, AVI	MPEG, SMIL	MPEG, Quicktime, Real Video
Sound	MP3, MPEG, Quicktime	MPEG, Ogg Vorbis	MP3, MPEG, Ogg Vorbis
Multimedia	SMIL, Macro Media Director, Assymetric Toolbook. All applications must be accompanied by full documentation and copies of native files, whether digital image, text, video, sound or database.	SMIL	SMIL, Native image, video, sound and delimited ASCII for database files.

Texts	ASCII text, RTF, HTML, PDF, Postscript, LaTeX, ODA, SGML, TeX, Word*, WordPerfect*, XML	ASCII text, HTML, XML	ASCII text, HTML, PDF, SGML, Word, XML
Databases	ASCII delimited text, MS Access, Oracle, Paradox, DBF	ASCII delimited text, XML	ASCII delimited text, XML
Spreadsheets	ASCII delimited text, MS Excel, Lotus 123, Quattro Pro	ASCII delimited text, XML	ASCII delimited text, XML
Geophysics	Contours, Geoplot, plain text	AGF, plain text	AGF, plain text

Table 2: AHDS delivery, preservation and dissemination file formats

Depositors are recommended to contact the AHDS centre for their discipline to confirm the file formats and versions of software packages which are currently accepted.

6.5.2 Contacting the AHDS

The Arts and Humanities Data Service has five centres devoted to archaeology, the visual arts, the performing arts, history and literature, languages and linguistics. Researchers who wish to discuss depositing virtual reality should contact the centre appropriate to their subject discipline:

- AHDS Centre for Archaeology: http://ads.ahds.ac.uk
- AHDS Centre for the Visual Arts: http://vads.ahds.ac.uk
- AHDS Centre for Literature, Languages, and Linguistics: http://ota.ahds.ac.uk/
- AHDS Centre for the Performing Arts: http://www.pads.ahds.ac.uk
- AHDS Centre for History: http://hds.essex.ac.uk/

Section 7: Resource Discovery

7.1 ACCESS AND USE

Use of virtual reality is expanding rapidly, with applications in museums, the broadcast media, the arts and in tourism and the heritage industries. The Virtual Reality Case Study Library illustrates some examples of recent applications of virtual reality, some created as museum or gallery installations and others disseminated via the Internet. The audience for VR is diverse and growing. This section of the Guide will look at uses of metadata to help people to find, access and use virtual reality worlds.

Metadata for virtual reality is important because it:

- helps virtual reality developers to find out about work that has been undertaken using particular techniques and enables research into VR techniques
- enables subject specialists to consider methods of interpreting and reconstructing sites in virtual worlds
- enables museum and media professionals to evaluate methods of mediating and presenting information
- helps artists to explore the uses of virtual reality in their work
- supports uses of virtual reality models in teaching and learning in schools, universities and colleges.

Access and reuse of all kinds of data are important because they help to support their preservation. The more uses for a dataset, the greater its chance of surviving beyond the immediate life of a project.

7.2 METADATA AND VIRTUAL REALITY PROJECTS

7.2.1 Introduction to metadata

Metadata is often described as data about data. It is information that helps a user or system to organise, access and use a resource. Metadata may serve various roles including: cataloguing and archiving, preservation, resource discovery and content description. These roles are often combined. For example, a library catalogue contains information that helps librarians to manage their collections (such as accession dates and the identity of the donor) as well as information (such as author names, titles, subject classification and shelf location) that supports resource discovery for library users.

Metadata has long been used in computer-based information systems. It provides information from the data dictionaries and 'system catalogues' in database management systems necessary for systems to locate individual items. Such metadata also provides structural information to aid users in understanding the contents of the database.

To be useful in resource discovery on the Internet, metadata must comply with a standard that provides a common descriptive format for diverse resources. The Dublin Core is an example of a simple metadata framework for the outline description of a wide range of resource types. It comprises fifteen basic elements: title, creator, subject, description, publisher, contributor, date, type, format, identifier, source, language, relation, coverage (spatial and temporal) and rights (see Miller and Greenstein 1997; Wise and Miller 1997). These elements are sufficient for simple resource discovery tasks, but they are not intended for detailed descriptions of complex resources. There are, however, many specialised metadata schemes for describing more complex resources. For example, the various MPEG standards address both technical specifications and content description.

Dublin Core descriptions are sufficient to determine that a resource deserves closer inspection, just as the author, title and abstract may suggest that a paper or book is worth reading. However, just as we read the arguments presented in a written work and follow up its references, we also need ways of examining the structure, content, sources and methods behind a virtual reality presentation. For virtual reality archives both basic descriptive (Dublin Core) metadata and more detailed descriptive metadata may be required.

7.2.2 Core metadata

The Arts and Humanities Data Service has been instrumental in promoting and exploiting the use of Dublin Core metadata to describe and enable access to digital resources. For example, the ADS catalogue currently contains about 400,000 metadata records describing resources that are either archived by the ADS or held by other institutions. A web-interface, ArcHSearch, provides ways for users to search this metadata.

The following is an example of a metadata record based on a virtual reality project for Canterbury Museum.

Information type	Scope note
Title	Quest for Canterbury's lost Roman temple
Creator	Nick Ryan, University of Kent http://www.cs.ukc.ac.uk/people/staff/ nsr/index.html
Subject.discipline	Archaeology
Subject.type	Temple
Subject.period	Roman
Subject.type	Excavation
Subject.type	Reconstruction
Description	A multimedia presentation for Canterbury Museums
Publisher	Canterbury Museums http://www.cs.ukc.ac.uk
Depositor	Nick Ryan, University of Kent

Date	20 October 1998 – 21 February 2001
Type	Interactive resource
Format	application/vnd.ms-powerpoint
Identifier	Not given
Source	Plans and excavation data
Language	English
Relation.archive	Plans and excavation data, Canterbury Museums 1980–2001
Coverage	Canterbury
Coverage. administrative area	Kent
Coverage.country	England
Rights.copyright	Museum display: Canterbury Museums 2001
Rights.copyright	Plans and excavation data: Canterbury Museums 1980–2001
Rights.copyright	Computer model: Nick Ryan 1998–2001

Table 3: AHDS Core metadata for virtual reality models

Relationship to the resource components

Virtual reality worlds are multimedia resources that comprise a number of component elements (program files, sound clips, images and so on). For such resources, the relation element of the Dublin Core metadata set may be used to record the component elements as follows:

Information type	Scope note
Relation.Has part	Video clip – Aerial fly in
Relation.referenced by	http://www.cs.ukc.ac.uk/people/staff/ nsr/va/des/fly_in.html

Table 4: Metadata describing the resource components

These fields can be repeated to identify all of the elements in a virtual reality application. However, it may be necessary to provide a core metadata record for each component resource and then gather them together in a collection level metadata record. In some cases the component elements may be maintained by different individuals or organisations. In such cases, the owner of each component would be responsible for maintaining its metadata. The project manager would be responsible for maintaining collection level metadata.

7.2.3 Educational metadata

Many virtual reality resources are developed for an educational audience. The Dublin Core education working group has proposed that the following additional metadata elements may be recorded for such resources:

Information type	Scope note
Audience	Museum visitors
Mediator	Canterbury Museums
Education standard	This resource does not conform to an established education or training standard
Interactivity type	Point and click presentation
Interactivity level	Low
Typical learning time	30 minutes

Table 5: Dublin core educational metadata elements

Appendix 1: Author Biographies

Tony Austin is the Technical Manager for the Archaeology Data Service. Tony is responsible for the development of ADS hardware and software systems. He is also responsible for the archiving of digital resources for the ADS collections and for developing the on-line user interface to these resources.

Rachael Beach is now based at Oxford Brookes University and was formerly a researcher with the Networked Virtual Reality Centre at the University of Teesside. She was involved in running the Building Babel II workshop and has been responsible for supporting creative 3-D computing in the Art and Design Community.

Aaron Bergstrom is the Multimedia and Digitizing Manager at the Archaeology Technologies Laboratory of North Dakota State University. He is a 3-D artist and programmer who is responsible for developing the DANA 3D multimedia viewers for the Archaeology Technologies Laboratory.

Sally Exon was Computing Research Officer at the University of Birmingham's Field Archaeology Unit where she was involved in developing the Virtual Wroxeter Roman Fortress. She continues to work in computing at the University.

Marc Fabri is an Associate Senior Lecturer at Leeds Metropolitan University where he is a member of the Intelligent Systems and Learning Environments Research Group.

Kate Fernie is a Research Officer with the Archaeology Data Service. She came to the ADS on secondment from English Heritage with experience of working with historic environment information systems and of writing and editing technical literature.

Michael Gerhard is an Associate Senior Lecturer at Leeds Metropolitan University where he is a member of the Intelligent Systems and Learning Environments Research Group. He has been involved in the project to develop CyberAxis – the collaborative virtual community for British Art.

Catherine Grout is the Multimedia Collections Manager for the Joint Information Systems Committee and was formerly the Manager of the Visual Arts Data Service.

Stuart Jeffrey is a research student at the University of Glasgow with research interests in 3-D scanning and visualisation of the historic environment.

Mike Pringle works for English Heritage in an advisory role in matters relating to the Internet and New Media technologies. He came to English Heritage, after studying for his PhD at the Department of Informatics and Simulation at the Royal Military College of Science, to develop PastScape – a virtual reality interface to heritage information.

Julian Richards is Director of the Archaeology Data Service. He is a specialist in the archaeology of Anglo-Saxon and Viking Age England and a leading expert on computer applications in archaeology.

Damian Robinson is currently a British Academy Postdoctoral Fellow in Roman Archaeology at the University of Bradford. He formerly worked for the Archaeology Data Service as Collections Development Manager, where he researched issues surrounding digital preservation and publication of archaeological data.

Nick Ryan is a lecturer in Computer Science at the University of Kent and his research interests include visualisation of ancient environments and documenting virtual reality. He developed the multimedia presentation 'Quest for Canterbury's lost Roman Museum' for Canterbury Museums.

Melissa Terras is based at the Department of Engineering Science and the Centre for the Study of Ancient Documents at Oxford University. She has a research interest in the use of 3-D virtual environments in museum education.

Case Study Authors

Kate Allen studied at the Exeter College of Art and Design, at the Akademie van Beeldande Kunsten in Rotterdam and at the Chelsea School of Art, London where she completed her MA in Sculpture. In 1988 Kate was awarded a fellowship in sculpture at the Gloucestershire College of Art and Technology. This was followed in 1990 by a scholarship in sculpture at the British School in Rome. In 1996 Kate registered as a PhD research student at Wolverhampton University to investigate the relationship between 'virtual' and 'real' sculpture and was awarded her doctorate in June 2002. A visiting lecturer at Chelsea College of Art & Design from 1993 and Reading University from 2000, she continues to exhibit 'real' and 'virtual' sculpture; her latest piece for Birmingham City Council will be launched in 2003.

Clive Fencott is a lecturer in the School of Computing and Mathematics at the University of Teesside. As part of his ongoing research he has constructed a VRML model of the Cliff Lift at Saltburn in the north east of England.

Learning Sites (Eben Gay, Geoffrey Kornfeld, Richard Morse, and Donald H. Sanders) is an American company which specialises in developing archaeological visualisations for interactive education and research. Learning Sites (http://www.learningsites.com/) developed the 'Northwest Palace of Ashur-nasir-pal II, Nimrud' and the educational package 'Ancient Greece: Town and Country'.

Anthony McCall Educated in England and resident in New York since 1973, Anthony McCall heads the Internet-based design practice, Narrative Rooms, LLC, (http://www.narrativerooms. com/) which designed and produced the pioneering on-line 3-D exhibition Brancusi's Mademoiselle Pogany, launched by Philadelphia Museum of Art in June 1998. Anthony McCall wrote the paper 'Visitors Online: Designing a Virtual Art Museum' (1996), which set out to explore some of the implications of this emerging on-line world.

Appendix 2: Documentation checklist

A2.1 INTRODUCTION

As discussed throughout this Guide, in order to support later reuse it is necessary to record information during the project in addition to creating a virtual reality model.

A2.2 CORE METADATA FOR THE AHDS

The metadata requested by the AHDS for digital archiving, listed below, is based on Dublin Core Metadata and is used for resource recovery through the AHDS on-line catalogues.

Resource discovery metadata is the index-level information that is used by gateways through which users seek archival material. 'Metadata', put simply, is data which describes the original information. The Arts and Humanities Data Service, along with a growing number of organisations around the world, advocates the use of the *Dublin Core* metadata set. This comprises a series of fifteen broad categories or elements, each of which is *optional* or may be *repeated* as many times as required. The elements may also be *refined* through the use of a developing set of sub-elements. The use of the Dublin Core within the Archaeology Data Service is discussed further elsewhere (Miller and Greenstein 1997; Wise and Miller 1997), and the current element definitions laid down across the Dublin Core community are available on the Internet at http://purl.oclc.org/metadata/dublin_core_elements.

Information type	Scope note
Title	The name of the bubble world, panorama or virtual reality model.
Creator	The name(s), address(es) and roles of the creator(s), compiler(s), funding agencies, or other bodies or people intellectually responsible for the model.
Subject.type	Keywords indexing the subject content of the model. If possible these can be drawn from existing documentation standards, e.g. for archaeology the English Heritage Thesaurus of Monument Types, the mda Archaeological Objects Thesaurus. If a local documentation standard is used a copy should be included with the dataset.
Subject.period	The time period covered by the virtual reality model.

Description	A brief summary of the main aims and objectives of the project for which the model was developed and a summary description of the model itself.
Publisher	List details of any organisation that has published the model including the URL of on-line resources.
Depositor	The name, address and role of the organisation or individual(s) who deposited the data related to the virtual reality model.
Date	The dates of the first and last day of the virtual reality modelling project.
Type	The type of resource, e.g. three-dimensional model, interactive resource, collaborative virtual environment.
Format	The data format of the resource, e.g. VRML 97.
Identifier	The identification number/code used internally for the project.
Source	References to the original material from which the model was derived in whole or in part from published or unpublished sources, whether printed or digital.
Language	An indication of the program language(s) in which interactions take place in the virtual reality model, e.g. Javascript.
Relation.archive	References to the storage location of archives associated with the creation of the model. This element includes the name and address of the archive repository, a description of the archive and details of how it is identified within the repository (e.g. by file name or accession number).
Relation.Has part	References to the component elements of the virtual reality model (e.g. image files, sound clips, video clips etc.) including details of their storage location, file names etc.
Coverage	Where the model relates to a real world location, give the current and contemporary name(s) of the country, region, county, town or village covered by the model and map coordinates (e.g. in the UK national grid).
Rights.copyright	A description of any known copyrights held in respect of the virtual reality model, its component elements or in its source materials.

Table 6: ADS Core metadata for virtual reality models

A2.3 A CHECK-LIST FOR DOCUMENTING VIRTUAL REALITY MODELS

This section offers a check-list of all the information recommended in this Guide indexed against the metadata requested by the AHDS. This is intended to assist project managers in compiling a summary of the project documentation that should be provided with a virtual reality archive.

Type	For AHDS
Project documentation (Section 5.2)	Y
Title	Y
Creator	Y
Subject •• Subject discipline •• Subject type •• Subject period	Y
Description	Y
Publisher •• Funding body •• Client •• Depositor	Y
Date	Y
Type	Y
Format	Y
Identifier	Y
Source •• Primary archive	Y
Language	Y
Relation •• Archive •• Component data files •• Report •• Bibliographic references	Y
Coverage •• Country •• Contemporary or current placename •• Administrative area •• Map coordinates	Y

Rights •• Copyright	Y
Target Audience (Section 5.3)	Y
Audience	Y
Mediator	Y
Education Standard	Y
Interactivity type	Y
Interactivity level	
Typical Learning Time	
Application development (Section 5.4)	Y
Model type	Y
Application format	Y
Application specification	Y
Hardware platform	Y
Authoring tools	
3-D drawing tools	
3-D scanners	
Object libraries	
Animation scripts	Y
Sound clips	Y
Images	Y
Real-world objects	Y
Interpretative objects	Y
Delivery platform (Section 5.5)	Y
Operating system	Y
Browser	Y
Plug-in/viewer	Y
Scripting language	Y
Hardware system	Y
Network connection	
Reporting project outcomes (Section 5.6)	Y
Report title and reference number	Y
Report author	Y
Report holder	
Report summary	Y
Description of archive (Section 5.7)	Y
List of all file names	Y

Explanation of codes used in file names	Y
Description of file formats	Y
List of codes used in files	Y
Date of last modification	Y

Table 7: A documentation checklist

Appendix 3: Information standards

Art and Architecture Thesaurus – produced by the Getty Information Institute and searchable over the World Wide Web. An invaluable tool for standardised description of material culture, architecture, and art in the western world from prehistory to the present. On-line: http://www.getty.edu/research/tools/vocabulary/aat/index.html.

English Heritage National Monuments Record Thesauri – available on-line and includes the Thesaurus of Monument Types, mda Archaeological Objects Thesaurus, Thesaurus of Building Materials and Maritime Thesauri. On-line: http://www.rchme.gov.uk/thesaurus/thes_splash.htm.

INSCRIPTION – is a collection of wordlists and thesauri which are maintained or recommended by FISH, the Forum on Information Standards in Heritage. Online: http://www.mda.org.uk/fish/inscript.htm.

Intra-governmental Group on Geographic Information, Principles of Good Metadata Management – An IGGI Working Group was established to prepare a best practice guide for the management of metadata and to increase the amount and currency of metadata held on the askGiraffe Data Locator. This Guide (Adobe Acrobat: 285kb) gives advice on the collection, management and dissemination of appropriate metadata and provides officials responsible for information handling with general guidance on managing metadata. On-line: http://www.iggi.gov.uk/achievements_deliverables/prinmeta.htm,

Rules for the Construction of Personal, Place and Corporate Names – The National Council on Archives 1997 guide to the recording of name information in archives. These rules include guidance on the use of non-current place names, and other issues of relevance to the archaeological community. On-line: http://www.hmc.gov.uk/nca/title.htm.

SPECTRUM: The UK Documentation Standard – Created by the mda in 1994 as a standard for documenting museum collections, its use is required for registration with the Museum and Galleries Commission. On-line: http://www.mda.org.uk/spectrum.htm.

TGN – The Thesaurus of Geographic Names is a project of the Getty Information Institute, aiming to create a powerful resource holding information on names of inhabited places, regions and geographic features both now and in the past. Importantly, TGN is holding these names within a hierarchy, such that it may be determined that a town lies *within* a country, that country *within* a continent, etc. On-line: http://www.getty.edu/research/tools/vocabulary/tgn/index.html.

Union List of Artist Names – produced by the Getty Information Institute and made available on-line, the ULAN offers a way to control the recording of names for over 100,000 artists and architects. On-line: http://www.getty.edu/research/tools/vocabulary/ulan/index.html.

Criteria for Evaluating Heritage Multimedia – The CIDOC Multimedia Working Group, chaired by Jennifer Trant, is working on a draft set of criteria for evaluating heritage multimedia resources. On-line: http://www.archimuse.com/cidoc/cidoc.mmwg.eval.crit.html.

JISC/TLTP Copyright Guidelines – This document, available as a pdf file, is targeted at the HE audience and covers a wide range of copyright issues in electronic media. A very useful document for reference. On-line: http://www.ukoln.ac.uk/services/elib/papers/other/jisc-tltp/jisc.pdf.

Simple Subject Headings – Designed by the mda for use in small museums, this tool helps control terminology used to classify museum collections. Four broad categories are included: community life, domestic and family life, personal life, and working life. This is a simplified subset of the mda's Social History and Industrial Classification. On-line: http://www.mda.org.uk/ssh_int.htm.

***word*HOARD** – The mda wordHOARD site provides connections to all manner of standards including the: International Association of Egyptologists' Multilingual Egyptological Thesarus, the Aquarelle Initiative and Term-IT. On-line: http://www.mda.org.uk/wrdhrd1.htm

BS 7666 – British Standard 7666 specifies the manner in which address information should be specified, and is likely to prove extremely important within Local Government and the Utilities. Archaeologically, it may prove most useful in the consistent provision of address information for Listed Buildings, etc.

FGDC – the United States Federal Geographic Data Committee's Content Standard for Digital Geospatial Metadata is, perhaps, the best known and established of geospatial data standards, the current version having been first released in 1994, and updated in 1998. This standard underpins much of the US Federal Government's work with geospatial data, and is also used by other collectors of spatial data. On-line: http://www.fgdc.gov/metadata/contstan.html.

Glossary

3-D space ball
These devices allow users to move or rotate 3-D models by moving a sensor ball, just as a standard mouse allows users to move a cursor on a computer screen.

Animation
In virtual reality, animation is the movement of an object or the viewpoint along a pre-determined path. Animation of the viewpoint or the user's view results in a fly-through or a guided tour. The animation may be repeated in an endless loop or have a set start and finish.

API
The application program interface (API) is a method by which an application program can make requests of the computer's operating system or of another application.

Avatar
The identity of a user of a CVE is represented by an avatar, which may be cartoon-like or a more realistic human figure. All simultaneous users of a CVE have visible avatars which reflect both an individual user's movement around the world and convey communications and emotions between different users.

   The term *avatar* derives from Hindu mythology. A deity called *Vishnu* is believed to have visited earth nine times to curb evil. For each visit, Vishnu took a different incarnation, called an avatar.

Backward compatibility
The ability for a computer application to read files which were created in a previous version of the software. A computer is said to be backward compatible if it can run the same software as the previous model of the computer.

Behaviours
Behaviours are program scripts that are attached to objects within VRML. The scripts cause an object to act in a certain way, for example a sphere may turn from red to green. This action may be triggered by a user of the world if the behaviour is attached to an event.

BMP
BMP is a standard image format in which image data is stored as a bitmap without applying any compression.

Browser
A Web browser is a piece of software which allows access to the World Wide Web. It interprets HTML, displaying the data in an easy-to-read format. There are two distinct types of browser – graphical and non-graphical/text only. Typical graphical browsers are Netscape

Navigator/Communicator and Microsoft Internet Explorer. An example of a text only browser is Lynx.

Bubble world
Environments created by manipulating 2-dimensional digital images and not by writing code which manipulates a computer's ability to display and render 3-dimensional geometry. Typically, as in Apple's Quicktime VR, a bubble world is made by taking digital photographs of the real place or space that is to be represented digitally. These are 'stitched' together into a 360 degree panorama. The user's viewpoint is in the centre of this panorama or 'bubble' and they can move around within its confines.

CAD
Computer Aided Design software is widely used by designers, surveyors, architects and others to produce 2-dimensional drawings and 3-dimensional models. For more information about CAD see the *CAD: Guide to Good Practice.*

Cave
Caves or sheds are projection-based virtual reality systems which use a system of display screens surrounding viewers to fill their field of vision.

CD-ROM
In computers, CD-ROM technology is both a format and system for recording, storing, and retrieving electronic information on a compact disk that is read using an optical drive. A disk can hold up to 600 megabytes of information.

Collaborative Virtual Environments (CVE)
This refers to an environment usually built in VRML (or some extension of VRML) that can be accessed by more than one user from more than one computer simultaneously. Users are made aware of each other's presence by the use of avatars and by the chat boxes which can be used to communicate with others.

Collision Detection (VR)
A program script that determines how close a user is to an object and stops their movement when they collide with the object.

CPU
The **C**entral **P**roccessing **U**nit is the component in a computer which performs operations on data. Data are input to the CPU, processed (according to the instructions in a program) and then output. Instructions can only be carried out one at a time, thus the speed of the processor affects the speed with which the computer works. Processor speed can range from 100 Megahertz (100 million cycles per second) to 2.8 Gigahertz (2.8 billion cycles per second). Desk-top computers generally incorporate slower processors than the workstations that are used by graphics designers.

DAT (Digital Audio Tape)
A helical-scan recording method initially developed to record CD-quality sounds on high-density audio tapes. It was quickly adapted for data storage applications. While DAT cartridges are all the same size (2.1 by 2.9 by 0.4 inches), the properties of the tape inside them differ. The smaller-capacity drives use tape cartridges that can store 1.3GB to 2GB of uncompressed

data, and they have typical transfer rates ranging from 183KB per second (KBps) to 366KBps. Their larger-capacity siblings support tape cartridges that store anywhere from 3GB to 4GB of uncompressed data, with typical transfer rates ranging from 366KBps to 510KBps. Many DAT drives offer some type of hardware-based data compression, which can significantly increase capacities and decrease transfer rates, depending on the type of data being stored.

Database

A generic term commonly used to describe a structured collection of data. Databases can take many forms including unstructured full text, images, maps, statistics or a mixture of data sources.

Dataglove

A glove that contains sensors which provide a means of controlling objects within the virtual world in direct response to movement of a user's hand.

Data model

The theoretical model by which data are structured. Common data models include relational, network, hierarchical and object-oriented. Data modelling is a methodology for structuring data for use in database systems.

Densitometers

A measuring device that registers the density of reflective or transparent materials.

DLT (Digital Linear Tape)

Digital Linear Tape Drive (DLT) provides a very fast (800 Kbytes per second) back-up to tape cartridges that hold either 20 gigabytes or 40 gigabytes of data and can be mounted in an automated library that holds enough cartridges to back up 5.2 terabytes of data

Dublin Core

A 15 field standard for metadata – or 'information about information'. Full details are available from: http://www.purl.oclc.org/metadata/dublin_core

DVD (Digital Versatile Disk)

This is an optical disk technology that is expected to replace the CD-ROM disk (as well as the audio compact disc) over the next few years. The digital versatile disk (DVD) holds 4.7 gigabytes of information on one of its two sides, or enough for a 133-minute movie. With two layers on each of its two sides, it will hold up to 17 gigabytes of video, audio, or other information.

Encryption

The conversion of data into a form, called a cipher, that secures against unauthorised access to data.

EPS

Encapsulated PostScript. An image-storage format that extends the PostScript page-description language to include images.

Events (VR)

An event is a program script which is attached to an object within VRML. The event triggers an action, or behaviour. For example, a script may cause a sphere to turn from red to green when a user performs a certain action. Events range from proximity sensors (users come within a certain distance to an object), to a timer (the user has been in a world for a specified amount of time) or a touch sensor (the user clicks on the object).

Facet (VR)
A facet is a planar surface of an object. Facets are generally triangular because triangles are always planar. Facets may be other shapes, as long as they are not warped in any direction and are truely planar.

Fishbowl VR
A term which refers to VR displayed and viewed on a personal, desk-top computer rather than on large projection screens or hemispheriums. The analogy is with watching a computer monitor as one would a fishbowl. Just as one expects to see a fish moving through time and space in a fishbowl, so fishbowl VR convinces us that we are watching and interacting with 3-D space 'in' the monitor. Also referred to as desk-top VR. See also Immersive/Non-immersive.

Fly-through
In virtual reality this is the movement of an object, or viewpoint, along a path that has been defined in a program script.

FTP
File Transfer Protocol. A common method for transferring files across the Internet.

GIF
Graphics Interchange Format. A bitmap graphics format from CompuServe which stores screen images economically and aims to maintain their correct colours even when transferred between different computers.

GIS
Geographic Information Systems are used to manage maps and other spatial data held in layers. GIS packages can hold data about the location and height of an object and increasingly are being used to produce two-and-a-half dimensional models of landscapes which can be animated. For more information about GIS see the *GIS: Guide to Good Practice* (online at: http://ads.ahds.ac.uk/project/goodguides/gis/).

Greyscale
The range of shades of grey in an image. The grey scales of scanners and terminals are determined by the number of greys, or steps between black and white, that they can recognise and reproduce.

Head-Mounted Display
A VR headset that restricts the user's vision to the VR environment since it covers the eyes. These are often used in immersive VR.

Hotspots (VR)
An identified point in a bubble world or VR environment that users can activate and cause a program script to execute an event.

ICC profiles
International Color Consortium colour standards. For further information see: http://www.color.org

Image compression
These are techniques which are used to reduce the size of digital image files. Lossless compression techniques, such as those used in the GIF and TIFF formats, retain all of the original image data while still reducing the overall file size. Lossy compression techniques, such as those used in the JPG format, compress the image file by removing image details (usually those details that the eye does not see very well) and thus losing some of the original data.

Immersive/Non-Immersive
The term immersive implies that an individual is experiencing VR either with a head-mounted display or else in some other manner, such as a hemispherium, which restricts their senses and reference to the real world. Non-immersive is generally referred to as fishbowl VR or desktop VR. From a qualitative point of view the different types of VR affect how the individual experiences the VR and how far they are convinced by the experience.

Interaction
This can be divided into low-level interaction and high-level interaction. Low-level interaction in the case of VR environments involves the user navigating around the environment and experiencing the space. High-level interaction is more complex and involves behaviours and events, that is objects act in a certain way when triggered by a user.

Internet Connection
This is the connection between a personal computer and the Internet and may be by cable modems, dsl modems, ISDN line, dial-up modem, satellite link, fixed wireless connection, etc. The type of connection affects the speed with which users can download files across the Internet. Speed can range from 56 K (kilobytes per second) in a dial-up modem to 3 mbps (Megabytes per second) with the cable modems and up to 100 mpbs with ISDN lines.

Intranet
A 'private' computer network, accessible only to particular persons, usually within a distinct organisation or institution. (As opposed to the Internet, which is a publicly accessible network.)

Interoperability
The ability of disparate computer systems to interact with one another, especially databases.

IP
Internet Protocol – one of the main protocols behind the working of the Internet.

ISO film speed
The standard for quoting photographic film speeds. It relates to the film's reactivity to light.

Java
Java is an object-oriented programming language that is designed to be portable across multiple platforms. It achieves this by using a 'virtual machine' known as the Java Runtime Environment (see below). Programs developed in Java, known as applets, are compiled for the JRE rather than for a specific operating system and thus can be run on any machine.

Java enabled browser

A web-browser that incorporates a JRE into its program is known as a Java enabled browser. Both Netscape and Internet Explorer incorporated versions of the JRE. But the pace of Java language development by Sun has meant that both Netscape and Microsoft have dropped JRE from their latest browsers. Users must now install a JRE plug-in from SUN Microsystems; this plug-in enables Java applets to be run within web-browsers or run directly from the user's computer.

Java Runtime Environment (JRE)

The Java Runtime Environment is a computer program which enables Java applets to run on different operating systems or web-browsers. Sun Microsystems develop the JRE for Sun, Linux and Windows operating systems. JREs are also being developed by freelance programmers for other operating systems, and Apple develops the Mac Java Runtime (MJR)

Javascript

This is Netscape's cross-platform scripting language, used for developing Internet applications.

JPEG

JPEG (Joint Photographic Experts Group) Image Format is a standard for variable level compressed images which are commonly used for display on Web pages. JPEG produces small file sizes using a lossy data compression technique.

Landscape

This refers to the orientation of an image. Landscape describes an image which is wider than it is tall. (An image that is taller than it is wide is referred to as 'portrait'.)

Level of Detail (LOD)

The level of detail visible in an object reduces with distance. In a virtual world, Level of Detail operations involve, for example, replacing a detailed object with a less detailed version at a set distance from a viewpoint or vice versa. Detailed surface textures may also be replaced by less detailed textures. Polygonal modellers are used to produce optimised versions of objects and textures which can be used in LOD operations.

Metadata

Metadata is often described as data about data. It is information that helps a user or system to organise, access and use a resource. Metadata may serve various roles, including cataloguing and archiving, resource discovery, technical and content description.

Multi-User Environment see Collaborative Virtual Environment

Node

This has two meanings in VR. In VRML, a node is a small piece of code which has a specific set of attributes. For example, the shape node can be either a sphere, a box, a cone or a cylinder as set out in the VRML specification. A node can also refer to a hot-spot in a bubble world that can be used to link together a series of such worlds. For example, a user can click on a hot-spot in a central bubble world and be transported into another bubble world. These might be a set of rooms in a museum, where users are able to move from room to room by clicking on hot-spots.

On-the-fly
Computer operations that develop or occur dynamically in 'real-time', rather than as the result of something that is statically predefined.

Open systems architecture
An architecture whose specifications are public. This includes officially approved standards as well as privately designed architectures whose specifications are made public by designers.

Operational specification
A definition of working parameters.

Optimisation (VR)
Optimisation is the process of improving the efficiency of a virtual world by removing any unnecessary facets that slow down the rendering of an object.

Photo-realistic
Representing an object 'as is', that is without any optical 'effects' etc. having been added.

Platform
A term that defines both the operating system of the computer and its hardware base, usually referring to the central processing unit.

Platform independent
Software or digital formats that can be used on any computer system regardless of the operating platform.

Plug-in
Plug-in applications are programs that can easily be installed and used as part of a web-browser, for example to view digital animations. See also Viewer.

PNG
Portable Network Graphics. Pronounced 'ping' The PNG format is intended to provide a portable, legally unencumbered, well-compressed, well-specified standard for lossless raster/bitmapped image files. Full details available from: http://www.eps.mcgill.ca/~steeve/PNG/png.html

Polygonal modeller (VR)
A virtual reality authoring tool or CAD software used to define facets creating 3-D objects. Polygonal modellers can also be used to edit objects or to optimise them. Some modellers automatically create less detailed versions of objects by reducing the number of facets. Optimisation tools remove any unnecessary facets that slow down the rendering of an object.

QTVR
Quick Time Virtual Reality is Apple's virtual reality format. In QTVR a panoramic image is projected onto the inside surface of a 'notional' cylinder or sphere and then viewed through an interactive window on the computer screen.

RAM
Random Access Memory, the part of a computer's memory where data are temporarily stored while being worked on.

Raster
A way of displaying spatial information as coloured grid cells. Also referred to as bitmap as effectively a map of bits is evident.

Ray tracing
A technique for adding realism to computer models by including variations in shade, colour intensity, and shadows that would be produced by having one or more light sources. Ray tracing software simulates the path of light rays as they would be absorbed or reflected by objects.

Real-time
If a computer responds in 'human time' this is considered to be real-time. For example, if a computer model moves approximately at the speed that users expect without being jerky or not rendering properly, it is considered to be real-time. As this is hard to achieve, a range of techniques is used to create an illusion of real-time movement. A different definition of real-time relates to the currency of information. This definition might apply if a user moved around a computer model and expected it to render and change instantaneously to display up-to-date information. An example might be a traffic map which users can access to see the speed of the traffic in the part of the city that they want to navigate.

Render
Adding realism to computer models, by for example applying a surface image to a geometrical frame.

Server
Computer that performs functions for other 'client' computers.

Stitching Program (VR)
A stitching program merges a set of images together to create a single large image without noticeable joins.

Storyboard
This is the process of making an outline of what a resource will look like before it is actually created. Storyboards are used by designers to organise the ideas and content used to convey a story. A high-level storyboard, in the form of a flow chart or in text, sets out events and identifies media requirements (such as photography, graphic design etc.). A graphical storyboard consists of sketches of virtual reality sequences, which may be accompanied by a script and a detailed description of how the user will interact with the content. Storyboards are modified throughout the design process.

Synthetic Environment
The military definition of a synthetic environment is a computer-based representation of the real world, usually a current or future battle space, within which any combination of 'players' may interact. The 'players' may be computer models, simulations, people or real equipment.

Textures
These are images which are applied to the surfaces of objects in virtual reality models to give the appearance of building materials or other surface details. Textures may be either photographs of real-world objects or simplified images that are created using drawing software.

Thumbnail

Low-resolution digital images, usually used for quick reference and linkage to a larger, higher quality image.

TIFF

Tagged **I**nterchangeable **F**ile Format/TIF (PC) or TIFF (Macintosh). A widely used graphic image format.

URL

Uniform Resource Locator. A standard addressing scheme used to locate or reference files on the Internet. Used in World Wide Web documents to locate other files. A URL gives the type of resource (scheme) being accessed (e.g. gopher, ftp) and the path to the file. The syntax used is: scheme://host.domain[:port]/path filename

VDU

Visual **D**isplay **U**nit – a computer monitor.

Vector

A geometric way of displaying spatial information as a series of points, lines and polygons.

Viewer (VR)

Viewers or plug-ins are software programs that are used to extend the capabilities of a browser or operating system. In the case of virtual reality, viewers to enable users to see models on desk-top computers.

Selected Bibliography

Allison, D., Wills, B., Hodges, L. F. and Wineman, J., 1997, 'Gorillas in the Bits', in *VRAIS'97*, The Proceedings of the IEEE Virtual Reality Annual International Symposium, pp. 69–77, IEEE Computer Society Press, California.

Arndt, S., Lukoschek, K. and Schumann, H., 1995, 'Design of a Visualization Support Tool for the Representation of Multi-dimensional Data Sets', in *Visualization in Scientific Computing* M. Göbel, H. Müller and U. Bodo (eds) Eurographics Association, Springer-Verlag Wien, New York, 190–203.

Barcelo, J. *et al.* (eds) 2000, *Virtual Reality in Archaeology*, BAR International Series 843.

Barfield, W., Hendrix, C. and Bystrom, K., 1997, 'Visualizing the structure of virtual objects using head tracked stereoscopic displays', in *VRAIS'97*, The Proceedings of the IEEE Virtual Reality Annual International Symposium, pp. 114–20, IEEE Computer Society Press, California.

Barthes, R., 1984, *Camera Lucida*, Flamingo.

Beach, R. and Birtles, L., 1999, 'Building Babel II: Construction in Virtual Environments', *Proceedings Computers in Art and Design Education '99 (CADE'99)*, University of Teesside, April 1999, pp. 211–14.

Beagrie, N. and Greenstein, D., 1998, *A Strategic Policy Framework for Creating and Preserving Digital Collections*, [on-line]: http://ahds.ac.uk/strategic.htm.

Blackmore, S., 1999, *The Meme Machine*, Oxford University Press.

Brooks, F., 1996, 'The Computer Scientist as Toolsmith', *Communications of the ACM* 39, No. 3, 61–68.

Carson, G. S., Puk, R. F. and Carey, R., 1999, *Developing the VRML 97 International Standard*, IEEE Computer Graphics and Applications, March–April 1999.

Cress, J. D., Hettinger, L. J., Cunningham, J. A., Riccio, G. E., McMillan, G. R. and Haas, M. W., 1997, 'An introduction of a direct vestibular display into a virtual environment', in *VRAIS'97*, The Proceedings of the IEEE Virtual Reality Annual International Symposium, pp. 80–86, IEEE Computer Society Press, California.

Earnshaw, R., 1997, *3D and Multimedia on the Information Superhighway*, IEEE Computer Graphics and Applications, pp. 30–31, March-April 1997.

Earnshaw, R., 1998, *Digital Media and Electronic Publishing*. London: Academic Press.

Edgar, R. and Salem, B., 1998, 'VRML Multi-user Environments', *Outline*, Issue 6, Autumn 1998, 32–4.

Eiteljorg, H., 2000, 'The Compelling Computer Image – a double-edged sword', *Internet Archaeology* 8 [on-line]: http://intarch.ac.uk/journal/issue8/eiteljorg_index.html.

Eiteljorg, H., Fernie, K., Huggett, J. and Robinson, D., 2002, *CAD: A Guide to Good Practice*, Arts and Humanities Data Service, [on-line]: http://ads.ahds.ac.uk/project/goodguides/cad/.

Ellis, S. R., 1996, 'Presence of Mind: A Reaction to Thomas Sheridan's "Further Musings on Psychophysics of Presence"', in *Presence* 5, No. 2.

Encarnacao, J. and Fruhauf, M., 1994, 'Global Information Visualization: the Visualization Challenge for the 21st Century', in *Scientific Visualization Advances and Changes* L. Rosenblum *et al* (eds), Academic Press.

Fencott, C., 1999, 'Content and Creativity in Virtual Environment Design' in proceedings of *Virtual Systems and Multimedia '99*, University of Abertay Dundee, Dundee, Scotland.

Fencott, C., 2001, 'Virtual Storytelling as Narrative Potential: Towards an Ecology of Narrative', in *Virtual Storytelling: Using Virtual Reality Technologies for Storytelling*, B. Balet, G. Subsol and P. Torguet (eds), LNCS 2197, Springer, Berlin.

Freeman, C. L., Webster, C. M. and Kirke, D. M., 1998, 'Exploring Social Structure Using Dynamic Three-dimensional Color Images', *Social Networks* 20, Part 2, 109–18.

Fuurht, B., Westwater, R. and Ice, J., 1998, *Multimedia Broadcasting over the Internet: Part 1*, IEEE Multimedia, Oct–Dec 1998.

Goodrun, A. A., 1994, 'Entertainment Technology and the Human-Computer Interface', *Bulletin of the American Society for Information Society* 21, Issue 1, 18–20, Oct/Nov 1994.

Haines, E.,1997, VRML 2.0 shading model applet, [on-line]: http://www.acm.org/tog/resources/applets/vrml/pellucid.html.

Herman, I., Delest, M. and Melancon, G., 1998, 'Tree Visualisation and Navigation Clues for Information Visualisation', *Computer Graphics Forum* 17, No. 2, 153–65.

Hibbard, W., 1999, *A Java and World Wide Web Implementation of VisAD*, [on-line]: http://www.ssec.wisc.edu/~billh/amsvisad.txt.

International Organisation for Standardization, (ISO)1997, *International Standard ISO/IEC 14772-1:1997: VRML 97*, [on-line]: http://www.iso.ch/iso/en/ISOOnline.frontpage.

Jeffrey, S., 2001, 'A Simple Technique For Visualising Three Dimensional Models in Landscape Contexts', *Internet Archaeology* 10, [on-line]: http://intarch.ac.uk/journal/issue10/jeffrey_index.html.

Jern, M. and Earnshaw, R. A., 1995, 'Interactive Real-Time Visualization Systems Using a Virtual Reality Paradigm', in *Visualization in Scientific Computing* M. Göbel, H. Müller and U. Bodo (eds) pp. 175–89, 1995, Eurographics Association, Springer-Verlag Wien, New York.

Kim, D., Richards, S. W. and Caudell, T. P., 1997, 'An optical tracker for augmented reality and wearable computers', in *VRAIS'97*, The Proceedings of the IEEE Virtual Reality Annual International Symposium, pp. 146–51, IEEE Computer Society Press, California.

Koller, D., Lindstrom, P., Ribarsky, W., Hodges, L., Faust, N. and Turner, G., 1995, 'Virtual GIS: A Real-time 3D Geographical Information System', in *Proceedings of IEEE Visualization '95 Conference*.

Litynski, D. M., Grabowski, M. and Wallace, W. A., 1997, 'The Relationship Between Three-dimensional Imaging and Group Decision Making: An Exploratory Study', in *IEEE Transactions on Systems, Man, and Cybernetics – Part A: Systems and Humans*, Vol. 27, No. 4, pp. 402–11, July 1997.

Loftin, R. B. and Kenney, P. J., 1993, *The Use of Virtual Environments for Training the Hubble Space Telescope Flight Team* [on-line]: http://www.vetl.uh.edu/Hubble/virtel.html.

Lombard, M. and Ditton, T., 1997, 'At the Heart of It All: The Concept of Telepresence', *Journal of Computer Mediated Communication* 3, No. 2, September 1997 [on-line]: http://jcmc.huji.ac.il/vol3/issue2/lombard.html.

Luecke, G. R. and Chai, Y-H., 1997, 'Contact Sensation in the Synthetic Environment Using the ISU Force Reflecting Exoskeleton', in *VRAIS'97*, The Proceedings of the IEEE Virtual Reality Annual International Symposium, pp. 192–98, IEEE Computer Society Press, California.

Macedonia, M. R. and Noll, S., 1997, 'A Transatlantic Research and Development Environment', *IEEE Computer Graphics and Applications*, pp. 76–82, March–April 1997.

Machover, C. and Tice, S. E., 1994, *Virtual Reality*, IEEE Computer Graphics & Applications, January, 1994.

Miller, P. and Richards, J., 1995, 'The Good, the Bad, and the Downright Misleading: Archaeological Adoption of Computer Visualization', in *Computer Applications and Quantitative Methods in Archaeology*, J. Huggett and N. Ryan (eds), BAR International Series 600, 19–22, Tempus Reparatum.

Miller, P. and Greenstein, D., 1997, *Discovering Online Resources Across the Humanities: a practical implementation of the Dublin Core*, Arts and Humanities Data Service and UK Office for Library and Information Networking.

Murray, J. H., 1997, *Hamlet on the Holodeck: The Future of Narrative in Cyberspace*, Free Press.

Nadeau, D. R., 1999, 'Building Virtual Worlds with VRML', *IEEE Computer Graphics and Applications*, March–April 1999.

Risch, J. S., May, R. A., Dowson, S. T. and Thomas, J. J., 1996, 'A Virtual Environment for Multimedia Intelligence Data Analysis', *IEEE Computer Graphics and Applications*, pp. 33–41, November, 1996.

Rosenblum, L., *et al.* (eds), 1994, *Scientific Visualization Advances and Changes*, Academic Press.

Rowley, J., 1998, 'Towards a Methodology for the Design of Multimedia Public Access Interfaces', *Journal of Information Sciences* 24(3), 155–66.

Sato, K., Shiratori, N., Nunokawa, H., Kusumi, T. and Syoichi, N., 1997, 'A User Interface Metaphor for Distributed Systems, Electronics and Communications in Japan Part III', *Fundamental Electronic Science* 80, Part 11, 82–93.

Serov, V. N., Spirov, A. V. and Samsonova, M. G., 1998, 'Graphical Interface to the Genetic Network Database GeNet', *Bioinformatics* 14, No. 6, 546–47, 1998.

Shawver, D. M., 1997, 'Virtual actors and avatars in a flexible user-determined-scenario environment', in *VRAIS'97*, The Proceedings of the IEEE Virtual Reality Annual International Symposium, pp. 170–79, IEEE Computer Society Press, California.

Spiney, L., 1998, 'I Had a Hunch ...', *New Scientist*, 5th September, 1998.

Stanney, K. M. and Salvendy, G., 1995, 'Information Visualization; Assisting Low Spatial Individuals with Information Access Tasks Through the Use of Visual Mediators', *Ergonomics* 38, No. 6, 1184–98.

Stone, R. J., 1998, *Virtual Reality: Definition, Technology and Selected Engineering Application Overview*, Mechatronics Forum Paper, May 1998.

Teylingen, R. van, Ribarsky, W. and Mast, C. van der, 1997, 'Virtual Data Visualizer', *IEEE Transactions on Visualization and Computer Graphics*, Vol. 3, No. 1, January–March 1997.

Vellon, M., Marple, K., Mitchell, D. and Drucker, S., 1999, *The Architecture of a Distributed Virtual Worlds System*, USENIX Conference Paper, [online]: http//:www.research.microsoft.com/vwg/papers/ oousenix.htm.

Warwick, K., Gray, J. and Roberts, D., 1993, *Virtual Reality in Engineering*, IEE, London.

Wheatley, P., 2001, 'Migration – a CAMiLEON discussion paper', *Ariadne* 29 [online]: http:// www.ariadne.ac.uk/issue29/camileon/.

Whitelock, D., Brna, P. and Holland, S., 1994, *What is the Value of Virtual Reality for Conceptual Learning? Towards a Theoretical Framework*, [on-line]: http://www.cbl.leeds.ac.uk/~paul/papers/vrpaper96/ VRpaper.html.

Wise, A. L. and Miller, P., 1997, 'Why Metadata Matters in Archaeology', *Internet Archaeology* 2 [on-line]: http://intarch.ac.uk/journal/issue2/wise_index.html.

Witmer, B. G., Bailey, J. H., Knerr, B. W. and Parsons, K. C., 1996, 'Virtual Spaces and Real World Places: Transfer of Route Knowledge', *International Journal of Human-Computer Studies* 45, 413–28.

USEFUL WEB RESOURCES

Bob Crispin's VRML Works: http://hiwaay.net/~crispen/vrml/
Dublin Core Education working group: http://dublincore.org/groups/education/
Guggenheim Variable Media Initiative: http://www.guggenheim.org/variablemedia/
Java Sun, organisational website: http://java.sun.com/
Java 3D Community Site: http://www.j3d.org/
Panoguide: http://www.panoguide.com
Superscape: http://www.superscape.com/
Web3D consortium: http://www.web3d.org/
Web3D repository – VRML resources: http://www.web3d.org/vrml/vrml.htm
Web 3D consortium – X3D specification: http://www.web3d.org/x3d.html
Web 3D repository – X3D resources: http://www.Web3D.org/vrml/x3d.htm

Virtual Reality Case Study Library

HOW TO LOCATE BROWSERS AND PLUG-INS

VRML
- Cosmo Player plug-in is the most popular VRML browser and is now distributed by Computer Associates International. On-line: http://www.cai.com/cosmo/.

Java 3D
- Java 2 Runtime Environment. On-line: http://java.sun.com/j2se/1.3/jre/
- Java 3D 1.2.1 – select the Java 3D for Windows runtime environment. On-line: http://java.sun.com/products/java-media/3D/download.html

Collaborative Virtual Environments
The following browsers and plug-ins are necessary to visit on-line CVEs on the World Wide Web. The respective websites give instructions on how to download and install the components.

- Blaxxun Contact plug-in to connect to blaxxun worlds. The installation procedure is fully guided and the plug-in installs itself as the default browser for VRML files. On-line: http://www.blaxxun.com/services/support/download/install.shtml.
- Active Worlds browser to connect to the Active Worlds server. The browser is a separate program that can be used in combination with a web browser. The browser updates itself automatically whenever a new version is available from the Active Worlds website. On-line: http://www.activeworlds.com/products/download.asp.
- Sony Community Place browser to connect to Sony Community Place Bureau servers. On-line: http://www.sony.co.jp/en/Products/CommunityPlace/down/index.html.
- Cosmo Player plug-in is the preferred VRML browser in combination with Netscape Navigator to connect to a DeepMatrix server. On-line: http://www.cai.com/cosmo/

'NETIQUETTE' IN COLLABORATIVE VIRTUAL ENVIRONMENTS

Standards of behaviour in Collaborative Virtual Environments may differ depending on what area of virtual space you're visiting, what your means of expression are, and if you are interacting with other 'real' people or simply leaving your traces in the virtual world. However, it can be said that these standards are not lower than in real life, and if you encounter an ethical dilemma in cyberspace, consult the code you would follow in real life.

Below are some guidelines for 'how to behave' in an on-line community:

- Avoid misunderstandings and always respect the human on the other end of the line. In CVEs, you probably lack in some or most of the communication means that are available in real-life conversations, such as facial expressions or gestures – channels that simplify and enrich face-to-face communication. Due to the limited set of channels in the virtual world, it is easy to misinterpret the messages of others, or to be misinterpreted
- However, there is nothing you can say that won't offend somebody! When being misunderstood, take the blame on yourself for being unclear, apologise, and say what you meant more clearly
- Be careful when using sarcasm and humour. Without face-to-face communications your joke may be viewed as criticism
- Be pleasant and polite. Don't use offensive language, and don't be confrontational for the sake of confrontation
- Know where you are in the virtual community. What's perfectly fine in one CVE can be pretty rude in another. When you enter a domain of cyberspace that's new to you, take a look around. Spend a while listening to the chat and get a sense of how the people who are already there act before you participate
- Keep in mind that because web-based CVEs are often globally accessible, a world can have inhabitants from many countries. Don't assume that they will understand a reference to TV, movies, pop culture, or current events in your country. Also, they might not understand geographical references that are local or national. If you must use them, explain them
- When being engaged in a serious conversation in an interactive CVE, don't expect instant responses to all your questions and contributions. And don't assume that all other inhabitants will agree with, or care about, your passionate arguments. It's easy to forget that other people have concerns other than yours
- Be aware that your actions can take up other people's bandwidth and rendering time. Use gestures and animations of your avatar appropriately. Bear performance in mind when designing your own avatar.

And finally, be forgiving of other people's mistakes. When someone makes a mistake, whether it's a spelling error or a stupid question or an unnecessarily long answer, be kind about it and think twice before reacting. Having good manners yourself doesn't give you licence to correct everyone else.

Useful Netiquette links to follow up:

- The Net: User Guidelines and Netiquette at Florida Atlantic University, on-line: http://www.fau.edu/netiquette/
- Netiquette by Virginia Shaw, published by Albion Books, on-line: http://www.albion.com/netiquette/index.html

Case Study 1: Brancusi's Mademoiselle Pogany at the Philadelphia Museum of Art
Anthony McCall, Narrative Rooms

A navigable on-line exhibition built in VRML for the Philadelphia Museum of Art website to describe and set in context a series of sculptures that the artist Brancusi developed over a nineteen-year period.

Launched by the Philadelphia Museum of Art on June 1, 1998, the on-line exhibition 'Brancusi's Mlle Pogany' was written, designed and produced by Anthony McCall and Hank Graber of Narrative Rooms.

The exhibition can be found at: http://www.narrativerooms.com/pogany/vr/index_a.html

Brief

Narrative Rooms proposed the Brancusi project to the Philadelphia Museum of Art. Anthony McCall developed the curatorial concept and the presentational aesthetic simultaneously and then worked with his colleague, Hank Graber, to realise it. Throughout the process the Philadelphia Museum of Art were interested and supportive observers, providing Narrative Rooms with generous access to their Brancusi collection but remaining firmly on the sidelines. However, when shown the finished result the Museum were extremely enthusiastic and launched the project officially on their site. McCall recommends a small team for this type of project with the active participation of a Senior Curator.

VR and the Museum

The digital presence of a museum on the World Wide Web can be fraught with difficulty. Planning has to take account of the fact that the digital museum must cater to different needs. A user may access the site merely to find out opening times before visiting the real museum. By contrast, another user may live many thousands of miles away and never visit the real museum but instead hope that the virtual museum has enough content to satisfy other curiosities. Spending large amounts of time and money replicating a site which uses the real architecture and look of the museum for its digital presence is irrelevant to the remote user.

In seeking a literal representation of physical reality some virtual reality designers look to architecture for inspiration. As a design discipline, architecture is driven by the constraints of physics and the manner in which people interact with their environment and each other in the physical world. On-line information environments will need to draw on these real architectural worlds for images and metaphors, but it makes very little sense to attempt to build "realistic" virtual replicas of real architectural spaces. Designers would be better served looking to theater, where narrative drives design, and visual references to the real world are expected to be interpretive rather than literal. The imagery of a theatrical set, for instance, may isolate a certain image, render it in exaggerated scale and give it a specific emotional coloring. This iconographic approach would also suit the small scale of the online experience, as well as being a practical response to the limitations imposed by bandwidth.

Many virtual museum authors seem to forget that whilst their building is an important part of their presence in the physical world, the museum actually exists as a concept to collect objects and create interpretive displays around them. Too often virtual museums are architectural rather than interpretive or narrative. Narrative Rooms' Mlle Pogany does not replicate a real room to the point of photographic reality. Instead it concentrates on displaying sculptures and related items in a 'suggested' reality. Only two architectural metaphors (inlaid floor and imposing doorways) are needed to situate the user firmly within a museum environment. Thus situated, the user can concentrate on the exhibits themselves.

What also differentiates the Mlle Pogany exhibition from other digital museums is its use of content and conventions created exclusively for the Web and not transferred from the real museum directly. Too often digital museums show exhibitions that are taking place in the real museums. Whilst this is useful perhaps for the prospective real visitor, a virtual visitor need not be served up a secondhand version of a real exhibition. Photographs of a real exhibition, when condensed into the scale and medium of the Web, serve to tell the virtual visitor nothing about the objects displayed. For instance a photograph of a real exhibition will not show the information boards in sufficient detail for the virtual visitor to participate in any interpretation of objects. Likewise the objects or paintings themselves are unlikely to have been produced in any detail. The Mlle Pogany exhibition uses that museum metaphor of information boards but with thought for the change in medium and scale, reproducing them on an exaggerated scale. Likewise it is selective in the photographs and supporting information that it reproduces at a sensible scale. Much thought has gone into the narrative experience and the way that the technology can be used to bring this out.

The Interface

Narrative Rooms decided that the proprietary VRML CosmoPlayer interface was inappropriate for what they were trying to achieve. They found it complicated, offering the average web user too many options for movement, thus allowing too many opportunities to become lost. They felt that this caused the user to concentrate on navigating rather than the actual content of the VRML world. Acting on this they customised the interface, simplifying the navigation only to that available with the familiar mouse. The price paid for this (loss of ability to look down, look up, slide sideways etc.) was felt to be worth it, as the visitor gained concentration and engagement with the content.

The Dynamic Map

In many digital museums the map is a continually used metaphor, allowing the user to perhaps click on it and enter that part of the museum that the clicked part of the map represents. This map is much more navigational and sophisticated as it exists at the side of the VRML window and it continually updates to show the user where they are in the exhibition at any one time. Initially Narrative Rooms did not provide a map. However, after discussing the experience of going round this first (mapless) version of the exhibition with colleagues and friends, they discovered that there was a desire to understand where the user was in the exhibition: how far they had come, how much more was there; and in addition, some reported getting lost and being uncertain where to turn to get back on track. The map was a response to this, and has been found to satisfy all those problems: it tells the user not only where they are, but, importantly, what direction they are facing (this is particularly useful when the visitor has been walking around one of the sculptures and wishes afterwards to proceed in the right direction).

Such a juxtaposition of 2-D and 3-D touches on one of the fundamental questions about 3-D navigation on the Web: how successful is it?

One of the claims for 3-D is that it provides a more intuitive interface than 2-D because it is more like the real world. Anyone who has suffered through the experience of navigating the typical VRML world on-line can tell you that it is nothing like the physical world; neither is it "naturally" more intuitive than other media that have had time to develop and whose conventions and vocabulary have become a well understood part of the culture.

Technical Details

Following an extensive period of research, a tentative structure was developed in which the project was blocked out in the form of drawings, storyboards, and written texts. The first 3-D model was constructed in Alias Power Animator running on an SGI workstation and exported to Cosmo Worlds, a VRML (Virtual Reality Modeling Language) authoring tool, for viewing.

Once the editorial content and layout of the exhibition were fleshed out, the team went to the Philadelphia Museum of Art with their mobile digital studio to digitise the two sculptures that were to be the exhibition's focus. Architectural details within the museum were also photographed to provide a starting point for designing the architecture of the virtual space. A 3-D laser scanning process gradually built up a three-dimensional topographical map of the object. A colour video camera within the scanner recorded the textures and colours of the object's surfaces and combined that visual information with the topographical data. The process is harmless to the object being scanned since there is no physical contact.

Back in the studio, the 1,200,000 polygons that made up the dataset of each sculpture had to be reduced to the 900 or so polygons that can reasonably be displayed on a standard home PC. The editorial materials and the 2-D digital images brought back from the museum were put into an image editing program on a Macintosh computer to create the final images for the 3-D exhibition; these included the photographic and text panels, plus the floors and doorways that became the defining architectural elements of this virtual world.

With all the elements of the exhibition assembled and the VRML programming done, the file was sent to Radical Virtual Reality in the Netherlands. The Java programmers at Radical customised the navigation interface to Narrative Rooms' specifications. The first complete version of Brancusi's Mlle Pogany was ready.

Tests were then conducted, both internally and within the museum, to check that the assumptions made about ease of navigation held up and to assess how effectively the exhibition was telling its story. This useful process helped the team to make significant technical and editorial adjustments before uploading the exhibition to the Philadelphia Museum of Art web site.

The use of two very basic architectural metaphors, that of the doorway and the floor, to suggest the space of the exhibition certainly has an aesthetic justification, but it is also a practical response to the necessity of keeping files small. McCall states that his colleague Hank Graber is a 'genius' at compression! The entire exhibition has a file size of only 400 kilobits.

For more information see 'Making Mlle Pogany' at http://www.narrativerooms.com/brancusi/makingMP.html.

Case Study 2: Exorcising the Flesh
Kate Allen

This case study is a personal account by the artist, Kate Allen, of the ideas and processes involved in creating VRML art installations and of audience reactions to them.

I began to combine photographic representations with physical objects, manipulating fashion models' torsos digitally, commenting on how the use of digital media in magazines, for example, may manipulate our ideas of a 'true' representation of reality. I found it frustrating that physical objects I built would be photographed then scanned before they existed on the computer. I wanted to move from manipulating photographs digitally to creating an entirely digital image, exploring the notion that simulation can produce a mathematical reality that has no previous existence in the physical world. The use of 3-D modelling software gave me the chance to explore this within my existing sculptural practice. With difficulty I learnt and am still learning to use Autodesk 3D Studio computer modelling software. I could now construct objects digitally, but to accept them as 3-dimensional I needed to animate the objects I made. Presenting the object as an animated film dictated what the viewer would see. I felt distanced from the objects I had made digitally, I wanted to make the exploration of the objects more interactive or more of a physical experience than watching an animation. This led me to VRML, which I had experienced on the Internet. I found VRML allowed me to explore the image – I could navigate through objects, find landscapes, views I had no idea existed, hidden inside the objects I constructed. It also allowed me to experience the object I had constructed in my own way as I would in the physical world. My desire to keep a sense of the unknown, a method of keeping that sense of discovery through making was regained using VRML.

To create my VRML objects I build the models in Autodesk 3D Studio and use a plug-in to change the model into VRML, later a plug-in became standard with Autodesk 3D Studio Max. I can create animations made in 3D Studio Max and export them to VRML. I had some difficulty getting animations working initially, as when the code is exported from 3D Studio to VRML the node for animation is written as loop FALSE instead of TRUE so I have to change the code in Wordpad. I use Cosmo Player to view my VRML models and I mostly use the examine mode. For me the VRML model is one piece of a whole installation/performance. I have only used a tiny part of the capabilities I can access through 3D Studio Max as I try to use the technology within the context of the ideas I have rather than becoming seduced by the possibilities of the software.

I created an installation, Have Your Cake and Eat it (1996). A VRML model of a cake projected on to a real cake allowing participants in the interactive installation to explore under the surface of the cake before biting into it. The objects inside the VRML cake represent some of the more unpleasant surprises which one may encounter while eating: finding a hair in your

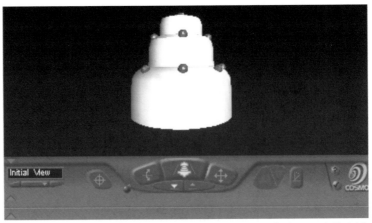

Figure 5: Images from the model 'Have Your Cake and Eat it'.

mouth, or seeing the chef with a runny cold sneezing over your dinner. It is probably better not to have this knowledge, but, because of the lure of the technology, or a fascination with the unpalatable, or simply because it is there, most of us will take a look. The interior of the virtual cake provides a new space: the 'inside' of the icing becoming an arctic landscape once one begins to travel through it. The VRML Cake has no matter; the newness of the technology makes more apparent the fact that the digital cake is nothing but a sign, and so is infinitely flexible and controllable, like our imagination. I was asked by several participants to tell if there really was a hair inside the cake they had eaten, another exclaimed that it was the first time they had ever been inside a glacé cherry!

Inspired by the texture of the VRML model and the sometimes difficult experience of navigating in virtual space, I built 'Lap Top Dog' (1997). This was a furry cover over the physical computer, where the viewer had to plunge their arms through furry sleeves to reach the mouse. The viewer could then manipulate a series of VRML images travelling through one into another etc. The objects based on internal and sexual organs, were nested inside each other, I wanted to create a sense of never-ending delving.

Figure 6: Images taken from the model 'Lap Top Dog'.

I wanted to make covers for the computers of people who were real expert programmers; I constructed them for people at the London Virtual Reality Group. I wanted the viewer to experience things getting hot, difficult but sexy enough to want to continue. I also wanted to continue the sense of the physical and the unknown.

My interest in exploring the interface of the VRML model with physical objects influenced the work 'Fat Free Fat' (1998), part of 'Exorcising the Flesh' commissioned by Walsall Museum and Art Gallery. The computer was housed inside an object, inspired from an individual pixel or one-calorie or piece of fat/flesh. A trackball mouse set into the front of the squashy sculpture manipulated the VRML object. The viewer looked through a tiny hole at the VRML object, a wobbling blancmange, and because of the restricted view the only way to view the object was to manipulate the model. The whole piece was impregnated with vanilla.

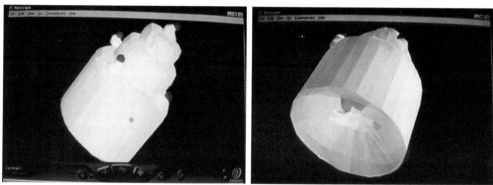

Figure 7: Images taken from the model 'Fat Free Fat'

Through experiencing audience reception of other computer works I began to get increasingly frustrated with the notion of interactivity. While watching people engage with my work I wanted to direct viewers to experience different parts of the piece. I noticed that quite regularly one person familiar with computers would interact with the piece while the rest of the audience looked on; through this person they experienced the work rather than interacting personally. This led me to the idea of performing with the VRML model. I thought I could act in a similar vein as a conductor of an orchestra, bringing out the best of a piece. I built a sort of ventriloquist dummy and created a performance work called 'Little Death' (1999–2000). This consists of a dummy of the computer animation star from Tomb Raider, Lara Croft, dressed in a wedding

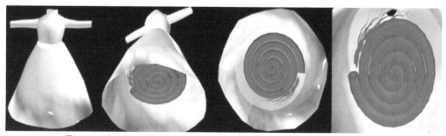

Figure 8: Images taken from the model 'Little Death'.

dress, a roller ball mouse set in her lap. Projected on to her are a series of VRML models – a wedding cake with cherries falling and a tongue licking the inside out of the VRML cake and a VRML wedding dress with Catherine Wheel internal organs spinning inside.

The model is manipulated by the roller ball mouse set in her dress. With my arm through her dress sleeve I can move the roller ball mouse and so move the objects. The wedding cake imagery with the cherries and tongue were a way of representing sexual pleasure. I have been interested in images of women created by men which is why I choose the Lara Croft character, built by a male programmer, a fantasy sex object for men. The rollerball mouse in Lara Croft's lap is moved to the instructions of the accompanying sound track of the music from the Sugar Plum Fairy, with directions spoken by a female saying 'left a bit up a bit right a bit'. The sound track and the movement of the mouse gradually becomes more excited until it almost reaches climax but not quite and the whole process begins again.

I am working on other pieces using VRML which has provided me with a sculptural experience using computer imaging, exploring the merging, confusion and manipulation of the 'real' and the 'virtual'.

Case Study 3: Virtual Saltburn by the sea: Creative content design for virtual environments
Clive Fencott, University of Teesside

Summary

This case study is based on a programme of ongoing research and a paper presented at the Virtual Systems and Multimedia conference in 1999 (see http://www-scm.tees.ac.uk/users/p.c.fencott/vsmm99/). This case study:

- Uses VRML for Heritage/Tourism websites; the model of the Cliff Lift was subsequently bought by the Saltburn Improvement Company
- Proposes a content model for the development of Virtual Environments (VEs) which advocates the planning of content in order to induce a feeling of presence for the user
- In this particular study, in contrast to the archaeological case studies also included in this Guide, the author advocates the inclusion of visual clues rather than accurate representations of objects
- Is based on over six years of experience in teaching VE design to both graduate and undergraduate students on both creative and technical courses.

Figure 9: Virtual Saltburn by the Sea

Virtual Saltburn by the sea is on-line at: http://www-scm.tees.ac.uk/users/p.c.fencott/Saltburn.

Introduction

What are we designing when we design a VE? Are we designing the scene graph, VRML nodes and so on? Are we designing chairs, tables, machines, creatures and so on? Are we trying to recreate reality or at least some imaginable reality? The answer to the first two is yes and no! The answer to the third is no! Eventually we will create chairs and tables and maybe creatures and whatever else but we we don't start by designing the scene graph let alone coding up VRML nodes. We are designing to communicate. We are designing to communicate the essence of things that visitors can complete for themselves in their own minds. We are also designing for 'the effective communication of causal interaction' (Ellis 1996) because VR is an interactive medium and we have to communicate its possibilities and the consequence of exercising those possibilities. We need to consider the initial stages of VE design at an appropriate level of abstraction that will help us create a model of perceptual content, which will in turn lead us to create a scene graph that the computer system can render as the sort of mediated environment that we want. That is what this paper is all about; the design of perceptual content. But first an example that we will use to illustrate the process.

The Cliff Lift at Saltburn

Saltburn by the Sea is a small Victorian seaside resort which was purpose built in the nineteenth century, having been simply a few seashore cottages used by fishermen and smugglers. As can seen from Figure 10, cliffs rise from the beach and to help the journey between beach and town (at the top of the cliff) a cliff lift was built in 1884. This consists of two small carriages running on an inclined tramway and connected by cables running round pulley wheels situated underneath the operators' hut at the top of the cliff. The motive power is water based. Each car has a tank

Figure 10: View of the cliff lift from the pier

underneath which can be filled with water. When some passengers are ready to travel, water is run into the tank of the car at the top of the tramway until its weight is greater than the car at the bottom. Letting the brake off a little establishes this fact. As the car at the top of the tramway moves down under the weight of the water and the passengers, so the car at the bottom is pulled up. When the short journey of 207 feet is completed the car at the bottom discharges the contents of its water tank ready to be pulled up on the next run.

Over the years, the seafront at Saltburn has changed little apart from the pier being shortened as a result of a storm in 1974. The buildings at the landward end of the pier, seen white with red geometric patterns in Figure 10, have been enlarged several times and now obscure the ticket office for the cliff lift which is at the foot of the tramway. Today the cliff lift at Saltburn is one of the oldest of its kind in the world and still carries some 70,000 people a year between town and beach and back.

Requirements and conceptual modelling

Requirements modelling parallels very closely the software engineering concept. One of the chief requirements is that purpose should be clearly established here.

Conceptual modelling is effectively the background research activity common to many design projects but in particular those with an aesthetic component. It is the gathering of materials, taking of photographs, sketches, plans, sound and video recordings, etc. It might also include the construction of mood boards as well as potential storyboards etc. Of course the world to be built might have no real-world counterpart and this will of necessity impact on the kinds of activities that might be undertaken here.

Conceptual modelling is where the VE builder gets to know the world to be built. It is also the stage at which major decisions concerning the type of content to be modelled are made.

The principal requirement for the Saltburn VE was that it had to appear fairly realistic but must be focused on trying to give visitors some sense of Saltburn as a seaside town with a strong Victorian heritage. The model also had to run with an acceptable frame rate on average home computers. Thus a few major activities for visitors to undertake, supported by a simple but evocative landscape, was the objective.

The town is quite small and separated from the seafront by steeply sloping, grassy cliffs. The seafront itself is relatively undeveloped by modern standards; there is a pub, a couple of gift shops, a surf hire shop (Yes! Saltburn is an important surfing centre), the pier – the only one in the north-east of England, an amusement arcade at the entrance to the pier and the cliff lift itself. Because of its relative solitude – even at the height of summer – and the high cliffs that rise immediately to the south there is also quite a lot of bird life. Capturing these in some way would offer something of the sense of Saltburn that is required.

One of the challenges in building this type of model is that it is a potentially unbounded, outdoor environment which would therefore constitute a greater test of the effectiveness of the design method. This is because the visitor has to be persuaded to remain in the area modelled rather than being constrained by walls and other physical barriers.

The decision taken was to model the cliff lift and its immediate environs at the foot of the cliff in some detail. The sea, beach, cliffs and cliff top would be kept very simple but modelled on a grand scale so as to make it difficult for visitors to find 'the end of the world'. In terms of interaction as a focus for visitors' attentions the following was decided upon:

- A working model of the cliff lift that can be operated by the visitor and that will also allow the visitor to ride it up and down
- Some sort of activity inside the amusement arcade
- Some bird life that is active in some sense on the beach
- Something to represent surfing would be desirable.

At this stage a sequence of prototype VRML models was developed which concentrated on investigating activities and interactions – using very simple geometry – that might be suitable for the final model.

Despite the fact that an apparently realistic model of a *real* place was being modelled, it soon became apparent that a whole series of choices needed to be confronted as a result of the necessary process of abstraction. In other words, the nature of perceptual reality as opposed to objective reality was being confronted. In the next sections, techniques for considering and modelling perceptual reality are investigated in the context of the Saltburn model.

Perceptual modelling

Perceptual modelling is the act of designing the perceptual opportunities and their inter-relationships. It is of course modelling the intended users' experience of the VE. Perceptual mapping (discussed below) techniques are used to build up a meta-narrative structure – analogous to the comprehensible labyrinth of Murray (Murray 1997) – of perceptual opportunities which are categorised according to the role they play in the planned scheme of possible user activity. Perceptual opportunities deal not only with conscious experience but also with unconscious experience, sureties, which deliver belief in the VE – perceptual realism in (Lombard and Ditton 1997) – irrespective of any real world counterpart. The existence and importance of unconscious experience is identified and discussed in e.g. Spiney (Spiney 1998) and Blackmore (Blackmore 1999).

The perceptual opportunities (PO) model of the content of VEs consists of a set of syntactic categories, which can be seen as attributes of any object or group of objects that might conceivably be placed in a VE (Fencott 1999). These attributes specify the way in which the object is intended to function in terms of both its ordinary, everyday meaning (denotation) and its special meaning as part of the interactive processes of the VE it will exist in (connotation). The syntactic categories into which POs can be characterised identify their role in achieving purpose and it is their planned interaction that gives us the overall structure we are looking for. We might thus see POs as a possible characterisation of the lexia – basic units of meaning – of virtual content rather than its scene graph representation. Figure 11 shows how the range of POs can be broken down into three principal forms that are briefly discussed below.

Figure 11: Perceptual Opportunities

Because of resource limitations, objects should only be placed in a VE if they provide a clearly identified perceptual opportunity integrated into the VE's perceptual map. They will then support the purpose of the world.

Surprises

The idea for surprises as perceptual opportunities came from the 'appropriately designed infidelities' of Whitelock *et al.* (Whitelock *et al*, 1994) who suggested them for emphasis in virtual worlds and thus to precipitate conscious learning. In other words, non-mundane details that are not predictable but they do arise, however surprisingly, from the logic of the space consciously accepted. Surprises, therefore, are designed to deliver the purpose of the VE by allowing visitors to accumulate conscious experience from which narratives can be constructed after the visit.

Surprises can be:

* implausible but beneficial
* totally plausible but unexpected

and there are three basic types:

* attractors
* connectors
* rewards.

Note that we have already identified four main foci for interaction in the conceptual modelling phase: the cliff lift, some bird life, something going on in the amusement arcade and something to do with surfing. We will see that these can all be modelled out of the basic set of surprises. We will also see that there are other large structures that are fundamental to VEs that POs can model as well. First, however, we will look at POs on the small scale in terms of units of interaction or agency.

Attractors

Attractors are perceptual opportunities which seek to draw the attention of a visitor to particular places or moments of interest in the VE. They will often be seen or heard from afar. Attractors can be:

* **Mysterious** – partially obscured/revealed objects, strange or unknown objects, both closed and open doors and doorways
* **Active** – movement, flashing lights, sounds changing pitch or volume
* **Alien** – objects that belong to another world, VE, or context altogether, 2-D maps, strange symbols to indicate the end of levels
* **Sensational** – objects which attract attention through non-visual senses, spatialised sounds, vibrations, smells etc.
* **Awesome** – large, famous, expansive, etc.
* **Dynamically configured** – objects that are relocated in space/time
* **Complex** – made up of a variety of attractors, perhaps seen at a distance
* etc. – this list is not exhaustive.

However, although many attractors rely on people's natural curiosity they are also directly related to other emotional involvements:

*   **Objects of desire** – objects that have some benign significance to the visitor and more particularly to the task at hand
*   **Objects of fear** – objects that have some malign significance to the visitor and to the task at hand.

Animation is a particularly successful form of attractor in that it makes things stand out because of our deep-rooted perceptual affinity for movement. However, attractors may be static and quite local. Doorways as both entrances and exits are examples of static attractors, as are partially obscured objects and localised sounds.

Examples of attractors in the cliff lift are:

*   The complex of buildings and the various sight lines of the horizon and promenade as seen from the initial startpoint (Figure 13), mysterious, awesome
*   Moving cars of the cliff lift 10 seconds after entry into the VE (also Figure 13), active
*   The oystercatchers on the beach in front of the pier from initial startpoint (Figure 12), active, mysterious
*   Open doorways of amusement arcade and ticket office, mysterious
*   Localised sounds in the amusement arcade, sensational.

All attractors rely on people's natural curiosity. Their prime purpose is to draw people into areas of conscious activity, called rewards, which are designed to deliver the main purpose of the VE.

Connectors

Connectors are the means by which visitors are helped to satisfy goals they have set themselves in responding to attractors. This could be through:

*   basic interaction technologies
*   information objects – maps, plans, signposts, etc.
*   on-line help
*   degraded reality – systematic removal of detail to deflect visitors back to areas of interest
*   and so on.

Examples of connectors in the cliff lift are:

*   the brake controls and the indicator signs over them which indicate which brake is active and what it does
*   the railings, steps and pathways act as connectors by providing sight lines to the main areas of interest
*   in addition, if the visitor travels away from the cliff lift along the promenade or cliff top, degraded reality is used to indicate they are leaving the main area of the VE. In this case the uprights on the railings disappear along with the presence of any other objects.

Rewards

Rewards come in various forms: local, peripatetic and dynamic, and should be designed to – as their name suggests – reward visitors for their efforts at interaction. The purpose of rewards is to deliver the specifically memorable experiences of the VE as well as ensuring that visitors linger appropriately from time to time as they move around the world. In virtual tourism, for instance, the longer visitors linger overall the more likely they are to find the virtual experience memorable and perhaps retain the desire actually to visit the place the VE is modelling.

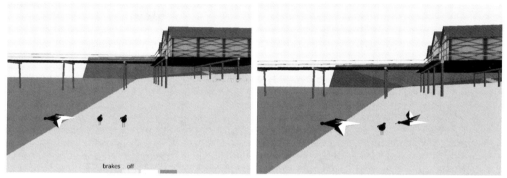

Figures 12a and 12b: Startled Oystercatchers

In the context of the cliff lift, the principal rewards are:

- Seeing the cliff lift cars change behaviour as a result of clicking the brake controls
- The oystercatchers flying away when the visitor gets too close to them
- Sounds changing as the visitor moves around inside the amusement arcade
- Encountering all the open doors having responded to the complex attractor of the initial viewpoint.

Perceptual mapping

Surprises should work together in patterns to form possible temporal orders or rewards and thus the coherent set of experiences that are intended to deliver the purpose of the world. These patterns are often called Perceptual Maps. The aim of perceptual mapping is to take the basic instances of attractor, connector, reward triples and compose them to build up a picture of the perceptual content of the VE as a whole. We can identify a number of larger structures:

- **Choice points** present the visitor with a choice of competing attractors or a choice of alternative goals for the same attractor. Murray (1997) identifies them as a major source of dramatic potential in VEs
- **Retainers** are groupings of surprises that constitute major sites of interest and/or interaction that seek to deliver the purpose of the VE as identified by requirements and conceptual modelling
- **Challenge points** can be as much conceptual as locational and are obstacles which have to be overcome in order to progress further in the VE
- **Routes** – implicit or explicit – draw visitors round a VE and seek to make sure that all major content is found and made use of.

Attractors should draw attention to sites of retainers and choice points etc. and, if properly designed, lead visitors around the world in a meaningful way using connectors. From the initial entry point the visitor is drawn to the complex of buildings in the centre of the frame. This is reinforced shortly afterwards (10 seconds) by the cliff lift cars moving up and down the tramway. As the visitor approaches the buildings, some steps up to a gap in the railings form a strong attractor, However, at about the same time the dark blobs near the shoreline become recognisable as birds and we have a classic choice point. Once on the promenade the competing attractors of the open doorways to the amusement arcade and ticket office form a powerful choice point.

Attractors may also themselves be retainers. Seen from a distance an animated object may act as an attractor (the cliff lift cars), but when experienced close up the object may be some sort of vehicle to ride in and control, thus becoming a retainer. Retainers are, in fact, patterns of attractors, connectors and retainers, which may be quite localised and, in effect, work as games. The oystercatchers avoiding visitors work in this fashion. They constitute a game in which the visitor chases them to find where they will go next. After a while the visitor realises that the birds always fly to one of three places on the beach and the little game is over, or at least poses no further surprises.

The attractor graph (Figure 14) shows the way POs can be used to construct routes through a VE. There are no challenge points as such in this VE.

Figure 13: View of virtual Saltburn

There are various means of documenting surprises as part of the design process:
- tables of suprises
- attractor graphs.

A partial table of surprises for the cliff lift looks like this:

Attractor	Connector	Retainer
Animated cars at distance	Brake controls	Operate cars from anywhere in world (outside cars) Peripatetic
Animated cars at distance	Pier and promenade railings, doorways of amusement arcade and partial views of ticket office etc. form axes to the car at the bottom of the cliff, alignment of pier and promenade	Ride up and down in the cliff lift Local

Unusual red and white pattern on amusement arcade, and partial views through entrance ways	Pier and promenade railings, alignment of pier and promenade form axes towards amusement arcade	Interactive soundscape inside amusement arcade Local
Shore birds on beach	Basic navigation controls	The bird always flies away when you get too close, you can chase it up and down the beach Local Example of a game as nested pattern of surprises
Partial views of path up/down (denoted by white railings up/down cliff)	Basic navigation controls	Climb up/down cliff with additional views of seafront etc. Local

Table 8: Surprises for the cliff lift

We can see that rows one and two constitute a choice point – two different goals from the same attractor. The representation of the shore birds attractor at different points on the beach documents a retainer. Retainers can be quite unprepossessing, for instance the path up/down the cliff. Such tables can be constructed in more detail by replacing retainers with individual rewards.

Another way of documenting perceptual mapping is by means of attractor graphs. These quite literally seek to establish the VE to be built as a comprehensible labyrinth of attractors as visible from other attractors. A partial example of such graph for the cliff lift is given in Figure 14 where the numbered nodes in the graph stand for the following attractors:

1   Buildings at foot of cliff, complex, mysterious
2   Cliff lift cars moving, active, mysterious
3   Shorebirds on beach
4   Ticket office with open doorways, complex, mysterious
5   Amusement arcade with open doorways, mysterious
6   Zigzag white lines of path up/down cliff, mysterious, complex.

The nodes in boxes represent retainers that could also be expanded as attractor graphs. The 'Operate cars' retainer has no links because it is peripatetic and can be undertaken almost anywhere and in conjunction with all the other retainers.

The advantage of attractor graphs is that they allow us to consider the main routes through a VE and the design of effective attractors to achieve this. It is also where major choice points should occur as these will be represented by nodes with more than one arrow leading away from them.

Perceptual mapping has much in common with the way painters arrange the composition of a work so as to catch the viewer's attention and lead it around the canvas in a particular way.

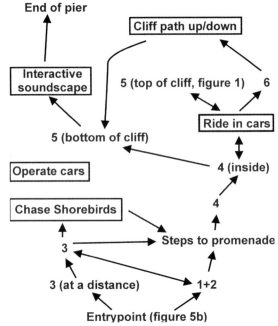

Figure 14: A partial attractor graph for virtual Saltburn

Although it is not possible to tell a story in a VE in the same way as in a film or TV programme, there is nevertheless an important narrative element to VEs which needs to be built into the design (Fencott 2001). This refers to the purposive accumulation of experience. This is more obvious in 3-D games or Virtual Training Environments (VTEs) such as the classic Hubble Space Telescope VTE used for training the flight team (Loftin and Kenney 1993).

Sureties

Sureties are mundane details that are somehow highly predictable – their attraction is their predictability. They arise directly from the architecture of the space and are concerned with the logic of the environment unconsciously accepted.

The following quote gives an insight from photography into the nature of sureties in VEs:
Hence the detail that interests me is not, or at least is not strictly, intentional, and probably must not be so, it occurs in the field of the photographed thing like a supplement that is at once inevitable and delightful. (Barthes 1984)

Sureties are about small things. Navigation for instance, lampposts, incidental furniture. This is because sureties for distance, as people would normally recognise them, are largely absent in VEs. This is also true for the scale of objects and one's own avatar. Space should not be static and sterile but dynamic and messy – we are used to the *real* world being like this so it helps if virtual ones are as well (VEs and mess/clutter don't however go naturally together). A useful aphorism is that in interacting with the real world we are trying to make sense of too much information but that in interacting with VEs we are trying to make sense of too little.

Sound is an important spatial surety in reality. It gives important information about the nature and scale of the space we are currently experiencing, i.e. small, large, inside, outside, etc. We are very susceptible to reflected sound as sureties in this sort of way. We are not very good at locating objects accurately in 3-D space based on the sounds emanating from them. The nature of sound in VEs means that sound can be used for atmosphere etc. but not as well for spatial and directional cues. This depends on the nature of the sound system itself being used, i.e. stereo, surround sound. Sureties are thus concerned with:

- vection
- ego scale
- perceptual noise
- distance
- limits.

Some examples drawn directly from the cliff lift are:
- sound of the cars moving over the metal tracks to reinforce travelling motion
- railings around the pier, the promenade and cliff tops to indicate avatar scale, vection and to suggest where to go etc. (It is worth noting the fact that none of the railings modelled bears more than a passing resemblance to those actually in place in Saltburn. Comparing Figure 13 above to the photograph of the pier should convince the reader of that. The actual and the virtual do not match. The important point is that there should be railings where railings are expected.)
- similarly the actual structure of the pier supports are quite complex but only the outer angled supports are modelled in the VE
- eaves on the amusement arcade and ticket office etc. which, as Figure 15 shows, do not resemble each other yet seem to have passed the scrutiny of visitors to date. Texture files have been scanned at very low resolution but this does not seem to have raised comments among the students who visited the site

Figures 15a and 15b: Real and virtual eaves

- eaves, railings, pier supports, etc. all add perceptual noise in addition to their primary design justifications.

Some elements have been either modelled very simply or not modelled at all:
- The sea is a single blue polygon that functions to signify 'sea' rather than model it with

graphical accuracy. This is both because of the computational overhead of a more detailed and/or dynamic representation and the risk that such a representation that did not work would draw attention to its inadequacies – see shocks below.

•   There are no cables connecting two cars together. Nobody has ever commented on this; what should be a shock goes unnoticed.

Sureties and surprises working together

Sureties and surprises in VEs work together in much the same way jokes do:

> My dog has no nose!
> How does he smell?
> Terrible!

The first two lines are unremarkable and mundane, sureties. The third line comes as a surprise but is plausible from the logic of the first two statements. Jokes all seem to be much like this – you set up an imagined and consistent, however fantastic, world and then give it a bizarre, implausible twist which must somehow be derivable from the former. Sureties and surprises in VEs work together, supporting each other and thus the virtuality they inhabit by seeking to catch and retain the attention of the visitor and thus maintain presence and belief. Sureties are the means by which a perceptual map is grounded, virtually, in a believable world.

In the context of the cliff lift we have:

•   the empty interior of the amusement arcade, filled with appropriate sounds but no objects, signals the use of the building

•   the fact that sounds, controls, animation, and vection come together in the cliff lift cars themselves establishes that they are the focus of the VE, particularly in relation to the partially modelled amusement arcade.

It is worth pointing out that many objects will provide a variety of perceptual opportunities as both surprises – quite possibly several depending on the context – and sureties.

Shocks

Shocks are not perceptual opportunities normally built into VEs but arise as by-products of the design and construction process. They give rise to perceptions that jar, that aren't received as expected in the established context of the VE. They draw attention to the mediated nature of the environment and thus undermine presence. Shocks are thus perceptual bugs which need to be actively sought out and eliminated.

In earlier versions of the cliff lift there were a variety of shocks such as the sea not being big enough so that visitors could see where it ended. The static nature of the sea as currently modelled is also something of a shock.

Conclusions

Despite being generally well received, there are unsolved problems with virtual Saltburn. Communicating the meaning and role of the lift control buttons at the bottom of the VE window has never been achieved. The surfing elements have never been incorporated due to the problem of making them perceptually convincing. The simplicity of the current representation of the sea is related to this. Virtual Saltburn is an empty place, devoid of the bustle of human activity that is the hallmark of the real Saltburn particularly on a hot summer's day. The increased power of even 'entry level' home computers means that many of these problems could now be solved if time permitted.

The perceptual opportunities model is not meant to be a prescriptive design method that tells designers what to do. It is a technique to help designers focus on achieving effective user participation and a coherent set of virtual experiences as a means of delivering purpose. It is also not the intention that designers should spend their time drawing complete attractor graphs or extensive tables of surprises. Once they are aware of the role of the various perceptual elements and the structures that can be built from them – choice points and retainers for instance – they can incorporate them into the general task of 'world building'. Over the last six years, student project work has shown this approach to be very effective in focusing minds onto the design of user involvement prior to 3-D graphical modelling and low-level scene graph construction. The result is that students build worlds that work effectively and coherently as virtual environments.

Case Study 4: Learning Sites
1. Northwest Palace of Ashur-nasir-pal II
and 2. Ancient Greece: Town and Country

Learning Sites (http://www.learningsites.com/) designs and develops interactive three-dimensional digital models that are based on archaeological evidence and reflect modern scholarship. Archaeological data is integrated with advanced computer graphics techniques to create virtual worlds that can be used for teaching, research, archaeological fieldwork, museum exhibitions and even tourism.

1: NORTHWEST PALACE OF ASHUR-NASIR-PAL II, NIMRUD, ASSYRIA

This model was produced and designed by Eben Gay, Geoffrey Kornfeld, Richard Morse, and Donald H. Sanders of Learning Sites, using archaeological data, interpretations and background text courtesy of Samuel M. Paley (State University of New York at Buffalo), Richard P. Sobolewski (Warsaw, Poland) and Alison B. Snyder (University of Oregon).

Description

The Northwest Palace of Ashur-nasir-pal II, at Nimrud lies in modern day Iraq. The palace was discovered around 1847 and was excavated on behalf of the British Museum in the 19th century and again in the 1950s. In the 1970s and 1980s there were further excavations by the Iraqis and the Polish Centre of Mediterranean Archaeology. Today the Palace is poorly preserved and is under threat from its natural environment, pollution, robbery and the political situation in the area.

The primary goal of this project is to bring together for the first time all the globally dispersed reliefs, sculpture, and related artefacts that have been removed from the Palace over the past 150 years and which has thus prevented scholars and the general public from fully appreciating this huge complex. It aims to document and publish a comprehensive, up-to-date report on the Palace, creating a multimedia and fully interactive scholarly research resource and teaching tool to be available both on DVD and the Internet with applications for colleges, universities, high schools, museums, and individual scholars. The project aims to address issues relating to the Palace, including, construction techniques, Assyrian politics and warfare, and artistic programs, as manifest in this first great monument of the Assyrian Empire and paradigm for all palaces to follow in the region. The presentation will take place within linked virtual worlds showing the Palace as it has been reconstructed and what elements of the building currently survive.

Learning Sites is in the process of creating an interactive 3-D computer model of the Northwest Palace, beginning with the Great Northern Courtyard and Throne Room suite. The project is using images scanned from photographs and drawings which are then applied to the digitally reconstructed walls to simulate the original positions of the carved bas-reliefs. The project aims to produce a very high-resolution virtual world which can be used for detailed archaeological analysis, incorporating newly produced photographs of the existing site and museum holdings and new research on the Palace architecture and sculptural program. Future developments may include innovations, such as the use of intelligent agents acting as virtual tour guides, programmed to respond to user queries on several different levels of scholarship.

Hardware and Software Used

Created and built to run on Pentium-based PCs, at least 166MHz or faster, with 48Mb RAM or more, Microsoft Windows 95 or higher, and Netscape Navigator 4.05 or higher, with the Cosmo VRML viewer plug-in. 3-D computer model created using DataCAD, textures and additional modelling with 3D Studio Max 2.5 and Photoshop, with associated plug-ins. VRML, Java, Javascript, and HTML programming in-house with standard tools.

2: ANCIENT GREECE: TOWN AND COUNTRY

This model was produced and designed by Eben Gay, Geoffrey Kornfeld, Richard Morse, Holly Raab, Donald H. Sanders and E.B. Sanders of Learning Sites, with archaeological data, interpretation and background information courtesy of the British School of Archaeology at Athens, Nicholas Cahill (University of Wisconsin) and John Ellis Jones (University of Wales).

Description and Objectives:

Ancient Greece: Town and country is a complete educational package for grades 6–12 and early college-level classes, using VR as the navigation and teaching tool to compare the House of Many Colours from Olynthus with the farmhouse at Vari, Attica. The package contains:

- virtual reconstructions of the houses
- VR models of the excavations
- interactive 3-D virtual models of artefacts from the two houses
- links to problem-solving tasks
- questions geared specifically to curriculum guidelines outlined in the United States
- a complete set of lesson plans and teacher workbooks linked to the worlds has also been included.

Further developments will include the addition of digital site models, more artefacts and furniture, interactive virtual characters performing daily life routines, and special VRML coding to allow users to see the virtual reconstruction of the houses 'materialise' from the ruins of the virtual excavation, and to do this space by space in order to compare the excavated evidence with our reconstructions.

Hardware and Software Used

Created and built to run on Pentium-based PCs, at least 166MHz or faster, with 48Mb RAM or more, Microsoft Windows 95 or higher, and Netscape Navigator 4.05 or higher, with the Cosmo VRML viewer plug-in. 3-D computer models created using AutoCAD, DataCAD, and Caligiari's TrueSpace; textures and additional modelling with 3D Studio Max 2.5 and Photoshop, with associated plug-ins. VRML, Java, Javascript, and HTML programming in-house with standard tools.

Case Study 5: Virtual Wroxeter: Roman fortress
Sally Exon, University of Birmingham

Virtual Wroxeter is a virtual reality model of Wroxeter Roman fortress in Shropshire. The model may be viewed on-line from http://www.bufau.bham.ac.uk/research/bt/default.htm or at the site museum at Wroxeter or Rowley's House Museum in Shrewsbury. It is also made available to schools on CD.

Wroxeter Virtual Fortress is a fully interactive educational package. Within the software, text is presented at a choice of three levels: UK national curriculum key stage 2 (age 9–11), key stage 3 (age 11–14) and an in-depth adult version. Background themes are introduced, including the Roman invasion of Britain, as well as topics interpreting the buildings of the fortress and life within the Roman army. The user explores the fortress using the computer mouse, moving around the reconstruction, 'walking' down streets, entering buildings or 'flying' over the fortress. Clicking on buildings displays text about them. There is also a guided tour option where the user is moved from building to building with text displayed at each part of the tour.

The Virtual Fortress was developed at the University of Birmingham Field Archaeology Unit (BUFAU) as part of a British Telecom sponsored project, Access to Archaeology. The project uses computer-based techniques to present data collected by the Wroxeter Hinterland Project to a wider audience (see http://www.bufau.bham.ac.uk/research/wh/base.html).

The reconstruction was created using Superscape's VRT software. Due to the size and complexity of the fortress plan, and to avoid splitting the reconstruction into several worlds, a variety of optimisation techniques was used to minimise the number of facets displayed at any one time. The reconstruction was to be web-deliverable so filesize was an important consideration.

Techniques employed in this way were:

- The inside and outside of buildings were built separately, the insides only being displayed as the user opens a door or enters a building
- The hospital was a very large building with many similar small wards off a central corridor. A small number of generic rooms were created. One of these is positioned in the correct place when the user opens a door to enter a ward. This helped to reduce filesize by removing the necessity for copies of rooms to exist behind each door
- The Software Development Kit was used to create an extension to VRT for performing Level of Detail (LOD) operations. VRT's own function (called distancing) calculates the distance of the viewpoint to the centre point of the object. This method is problematic if objects are very large or asymmetrical (such as fortress defences). In this situation the viewpoint may be close to an edge of an object but the distancing function calculates that the centre point is far away and swaps to a simpler version of the object. Splitting objects

up into many smaller parts and distancing on each would have increased the number of facets displayed and the file size to an unacceptable degree. The extension that was developed calculates the position of the viewpoint from the bounding box of the object allowing large asymmetric objects to be used in the world.

The fortress and pages decribing it are highly integrated. Interaction between the VR and HTML was implemented using Javascript, DHTML and stylesheet techniques.

The use of low-resolution PC-based virtual reality techniques allows this software to run without the need for expensive specialist hardware or software.

Case Study 6: Quest for Canterbury's lost Roman Temple and metadata case study Nick Ryan, University of Kent

Public presentations, whether on the Web, in museums or in the broadcast media, can be much more than a repackaging of earlier forms in a fashionable and spectacular medium. They do not need to be limited to telling a single story or to presenting one of many possible interpretations as an established 'truth'. Figure 16 shows an introductory frame from a multimedia presentation, 'Quest for Canterbury's lost Roman Temple', developed for Canterbury Museums. One of the purposes of this system is to show museum visitors how archaeologists can make significant inferences about the layout of a Roman town and the form of its buildings from very incomplete excavation evidence and a knowledge of comparable structures elsewhere. The presentation has been designed as the first part of a system to provide information about each of the main public building complexes in the city, although the current implementation concentrates on a postulated classical temple and its precinct. The other complexes (the forum, public baths and theatre) are shown only as still images with minimal descriptive text.

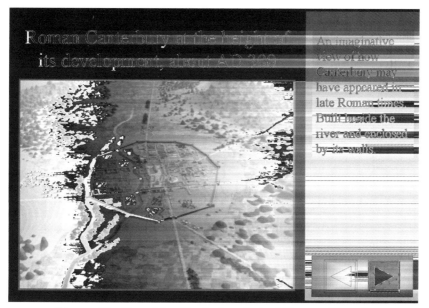

Figure 16: Quest for Canterbury's lost Roman Temple

The temple is thought to have been situated in a precinct adjacent to the forum and theatre complexes. A portico surrounding the precinct has been located in several excavation trenches, but there have been few opportunities to explore the enclosed area. Where this has been possible, evidence of an extensive courtyard surface has been found, often cut into by post-Roman structures. Materials that could come from a significant classical building have been found re-used or re-deposited in these later contexts within the precinct area, but the exact location of any temple remains unknown. However, there is anecdotal evidence to suggest that its foundations were observed, and possibly largely destroyed, during building work in the 1960s, before the establishment of a full-time excavation unit in the city.

The presentation runs on a touch-sensitive flat-panel device standing next to a display case containing finds from the area. These finds include a number of masonry fragments, mostly from one or more Corinthian columns, and a variety of marble and other stone wall-cladding materials. The presentation, therefore, seeks to breathe some life into these otherwise rather dull finds that might normally be labelled only with dry statements such as 'Masonry fragments, possibly from a Roman temple'.

Interaction is minimal. At a few points, the user can choose which building complex to visit, or make minor changes to their route through an otherwise fixed sequence of frames. Buttons are also provided to skip backwards or forwards if the user finds the default timing too fast or too slow. The latter is more likely because the default timing is aimed at the needs of younger and potentially slower readers.

Most of the display sequence uses still images and video fly-through or walk-through sequences of complex solid models of the city in its surrounding landscape (Figure 16) and individual buildings, together with textual annotation in a choice of three languages. Animations are used to overlay models on excavation plans and to place photographic images of the finds displayed in the adjacent case onto rendered solid model representations of Corinthian columns and capitals.

The simplicity of the system and the use of 'guided' or 'directed' sequences, rather than allowing users to roam freely around a virtual Roman Canterbury, was a deliberate design choice. Canterbury Roman Museum is small, yet it attracts large numbers of visitors and is popular with parties of children from local schools. The design of the museum itself follows a similar pattern, with visitors encouraged or constrained to follow particular routes through its various rooms. The undeniable attraction of interactive displays, particularly to younger visitors, led to the requirement that the maximum duration of the display should be tightly constrained to prevent queues forming.

At an early stage of the design process, VRML was considered as a means of implementing the three-dimensional models, and simple prototypes of the temple and theatre were shown at various 'open days' to give an impression of what future displays at the museum might look like. In the end, the need for directed sequences and the relatively poor image quality of most VRML renderers led to the decision to prepare all images and video sequences using a solid modelling and ray-traced rendering solution. Vue d'Esprit by e-on software was used both as a modelling and rendering tool.

Source data used in the model included a DEM derived from out-of-copyright map data, and known building plans which were derived from both published and unpublished excavations undertaken by Canterbury Archaeological Trust and other groups. Many building fragments were initially drawn using various CAD programs and exported as DXF or DWG files, whilst others came from earlier projects using other modelling and rendering tools such as POV-Ray. The remainder of the model was developed within Vue d'Esprit.

Despite the availability of several more sophisticated multimedia authoring packages, this system was developed using Microsoft PowerPoint, a medium more often associated with conference presentations and lecture slides. Apart from the author's familiarity with this package, this choice was made because of PowerPoint's simplicity and adequacy for the relatively simple requirements of the display system. Indeed, the system could have been realised using almost any of the authoring packages currently available.

One of the benefits of the early VRML-based prototypes was that they were intended to be viewed using an HTML browser, with an appropriate plug-in, and so might also be published on the Web. Although the system installed in the museum employed a conventional multimedia approach, the possibility of Web-based delivery was not wholly abandoned. The same content has been used to produce several, albeit incomplete and unpublished, versions of the system as test-beds for more recent approaches. One of these is discussed later.

The image in Figure 17 is a frame from the beginning of an initial video sequence that flies in from an aerial view to land in the centre of the town. The 3-D model of Roman Canterbury and its surrounding landscape is based, where possible, on archaeological and environmental evidence. However, in a model of this size, there are large areas for which no such evidence is available and in these the model elements are no more than informed speculation. Even where evidence from archaeological excavations is available, it is invariably partial. Excavations rarely recover complete building plans and structures rarely survive above floor level and are often represented only by robber trenches.

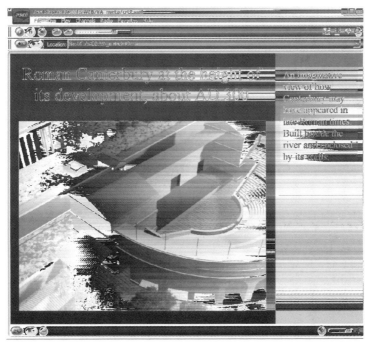

Figure 17: Video frame from the quest for Canterbury's lost Roman Temple

The model represents a composite of much of what is known of Canterbury during the first four centuries AD. It is intended to convey an impression of the town's appearance during much of this period, rather than to be a strictly accurate model of any particular date. The date of 300 in the title was chosen as a compromise to allow the late 3rd-century walls to be included, even though the building complexes that are the main focus of the presentation are mostly of 2nd-century origin. Indeed, some earlier buildings in the model may have gone out of use and have been demolished or replaced by later structures by this date. Most of the circuit of Canterbury's medieval walls remains standing today, although large sections are more recent rebuilds, and these are founded on the earlier Roman walls. These walls are, therefore, an important local landmark and their inclusion in the model helps viewers to orient themselves.

This model and the system by which it is presented to museum visitors are quite typical of current uses of Virtual Archaeology for public presentation. The presentation is a little unusual in that it seeks to convey an understanding of the incomplete nature of the archaeological evidence and the resulting uncertainty inherent in its interpretation. Other than this, however, it offers much the same benefits and suffers many of the same limitations as most similar systems in current use.

Information type	Scope note
Title	Quest for Canterbury's lost Roman temple
Creator	Nick Ryan, University of Kent http://www.cs.ukc.ac.uk/people/staff/nsr/index.html
Subject.discipline	Archaeology
Subject.type	Temple
Subject.period	Roman
Subject.type	Excavation
Subject.type	Reconstruction
Description	A multimedia presentation for Canterbury Museums
Publisher	Canterbury Museums http://www.cs.ukc.ac.uk
Depositor	Nick Ryan, University of Kent
Date	20 October 1998 – 21 February 2001
Type	Interactive resource
Format	application/vnd.ms-powerpoint
Identifier	Not given
Source	Plans and excavation data
Language	English
Relation.archive	Plans and excavation data, Canterbury Museums 1980–2001
Relation.Has part	Video clip – Aerial fly in
Relation.referenced by	http://www.cs.ukc.ac.uk/people/staff/nsr/va/des/fly_in.html
Coverage	Canterbury

Coverage. administrative area	Kent
Coverage.country	England
Rights.copyright	Museum display: Canterbury Museums 2001
Rights.copyright	Plans and excavation data: Canterbury Museums 1980–2001
Rights.copyright	Computer model: Nick Ryan 1998–2001
Audience	Museum visitors
Mediator	Canterbury Museums
Education Standard	This resource does not conform to an established education or training standard
Interactivity Type	Point and click presentation
Interactivity Level	Low
Typical Learning Time	30 minutes

Table 9: AHDS Core metadata for Canterbury Museums VR presentation

Case Study 7: CyberAxis
Michael Gerhard, Leeds Metropolitan University

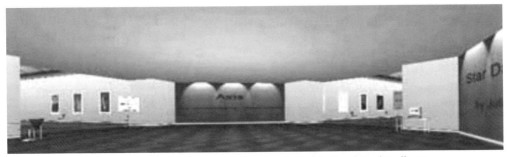

Figure 18: Reception area in the CyberAxis virtual gallery

The Intelligent Systems and Learning Environments (ISLE) Research Group (http://www.lmu.ac.uk/ies/comp/res_group3.html) at Leeds Metropolitan University (LMU) in co-operation with Axis, an organisation for information on contemporary visual artists in Britain, is developing a web-based, multi-user, virtual gallery. This joint project, named CyberAxis, has been granted financial support from the government's *Millennium Festival Fund for the Arts* and various other sources. The aim of the project is to increase the accessibility of contemporary British art by developing a 3-D virtual world as a new interface for the Axis database.

The Axis database has been established as the National Artist Register since 1994 and contains multimedia information on over 3,800 artists and 15,000 artworks. Traditionally, access to the Axis database is from a number of nationwide Axispoints. Axispoints have a 2-D, single-user, graphical interface. They are connected via the Axis Intranet and updated monthly, hence representing an up-to-date, but static, snapshot of information on contemporary art in Britain. The recent increase in accessibility and usability of the Web made it a promising platform for delivering the Axis database. The implementation of full web-access to the database was a first step towards increased accessibility.

Furthermore, new communications technologies can give access to visual arts in new ways and could be very effective in widening the audience for contemporary visual artists, particularly amongst young people. The recent rapid increase in 3-D multi-user virtual environment technologies available for the Web has the potential to offer a new kind of experience. As this technology is emerging, it is now possible to exhibit artworks of the Axis database in a virtual world; a web-based, multi-user, three-dimensional, virtual gallery. Within this shared space, avatars represent visitors and artists who can come together to spread information, share their views on art or simply chat.

CyberAxis went live on 14 June, to celebrate the Year 2000 – Year of the Artist – and Contemporary British Art, with a series of four exhibitions – Julie Mackies' exhibition 'Star Date 2000'; 'Landmarks' by Angela Edmonds with Teresa Pemberton in residence; Jenny Rumens and Sally Wallace's exhibition 'Exploring the Possibilities: an Eclectic Approach'; and the fourth exhibition 'Cabinet of Curiosities' by Chila Kumari Burman. All of these exhibitions can still be viewed.

Figure 19: Visitors represented by avatars in CyberAxis

Since the launch of CyberAxis it became apparent that technology was still a hurdle that not everybody was able to overcome. The download and installation of the VRML browser plug-in caused some problems on low-specification machines with slow Internet connections. However, visitor numbers have shown that the virtual gallery has been accepted and is highly appreciated by some. During the course of the four initial exhibitions the CyberAxis home page received 4,607 hits, 632 distinct visitors made their way to the interactive CyberAxis virtual gallery, engaging in some interesting debates that took place during scheduled artist residencies. A further 581 users have viewed the gallery in single-user mode. Overall, the feedback that Axis received was very positive and it has proved to be a step in the right direction.

Axis hosted events in the CyberAxis gallery during 2000 and 2001. Users of the Axis on-line database were invited to nominate their favourite artworks to be exhibited in the 'Common Curator' exhibition, which went live in CyberAxis in spring 2001. CyberAxis was hosted by Cybertown and was based on Blaxxun software (see Section 4.6).

The CyberAxis virtual gallery has now closed, although it is still possible to view previous exhibitions in single-user mode. More information on CyberAxis is available on-line at http://www.axisartists.org.uk/cyberaxis/default.html.

Case Study 8: Building Babel II
Rachael Beach

Abstract

This paper documents an experimental workshop that took place over three days in Sept 1998 at Coventry School of Art and Design (Beach and Birtles 1999). The workshop was an exploration of the issues surrounding the construction and use of CVEs by the art and design community, though our conclusions are relevant to the use of CVEs in all disciplines.

Building Babel II developed in response to our experiences of teaching VRML and to the problems of content creation and collaboration we perceived CVEs to have. This paper documents our attempts to solve these problems by integrating existing CVE technologies with art and design working practice.

Introduction

As a group we had previously undertaken some research into currently available collaborative multi-user spaces. In most cases we found that technologies such as Active Worlds and Superscape offered highly optimised and efficient 3-D, collaborative representations of reality and, in the case of Active Worlds in particular, some facility to alter these representations. It soon became clear to us, however, that for the art and design community such preconceptions of reality imposed an unacceptable limitation on creativity. Whilst we were under no illusion that the CVE had been implemented for virtual 'homesteading' and not for collaborative building of alternative realities by artists and designers, nonetheless we felt that these CVEs held great potential for meaningful collaboration if only we could provide the user with the ability to create their own objects and worlds. We also perceived that, in our experience, meaningful collaboration was not actually easy to achieve through the interfaces provided in existing CVEs, and that additional support would be required for our artists and designers, some of whom had little or no working knowledge of VRML when they put forward a working proposal to us.

Having considered these problems, we felt that this meaningful collaboration was best achieved by the use of existing CVE technology integrated into certain aspects of art and design practice. So we decided to adapt existing CVE technologies and integrate them into a workshop in which participants worked collaboratively in both real and virtual spaces that mimicked environments that they would already be familiar with. This would, we surmised, provide the support for participants which we felt they needed to deal with these technologies.

VRML and the CVE server

In terms of existing technology we made the decision to use a VRML-based CVE. From a technical perspective we chose VRML for its adaptability, stability, specification and the wide variety of resource material available. More importantly, however, it has been our experience that VRML appeals to artists and designers, because it can be manipulated outside of its imagined use to describe abstract concepts, forms and behaviours.

We considered Vnet, Blaxxun and Community Place as VRML CVE servers. Vnet initially seemed the ideal solution to our needs, allowing us to bring together both Macintosh and PC users. However Vnet requires a level of understanding of VRML and Java that we felt would shift the focus of the workshop to that of overcoming technical issues rather than of creating environments. We found that Sony's Community Place was simply lacking in support and developer information and so we settled on the Blaxxun client and server pair. Despite its complex interface involving a series of HTML frames and embedded files, Blaxxun offered a stable server which would allow participants to upload a standard VRML world which could immediately be shared. We did, however, sacrifice some stability and functionality, such as the facility to alter the environment while present in it, as is possible in Active Worlds. In addition Blaxxun has a wide selection of support material and examples available for the server.

Security and trust

By deciding to provide a high level of control for the participants, i.e. access to the server, we incurred the security risk of providing access to configuration files and scripts. With this amount of access it is a relatively simple task to halt the server. Our choice though was to trust that participants would not wish to jeopardise their work by making untried, frivolous or malicious changes. Anyone visiting the experiment from outside the working group was not given permission to write files.

Our trust in the participants and their trust in us was a fundamental consideration when planning the workshop. Since the workshop was intended to investigate collaboration, we conceived of this collaboration on two levels, both the 'real and 'virtual'. CVEs are generally investigated from the point of view of remote working and so collaboration in that sense is truly virtual, but we believed that our participants would benefit from meeting each other and from

Figure 20: The real space

meeting us. We hoped that providing this facility would encourage meaningful dialogue between participants and any trust and rapport that they formed would be transferred into the 'virtual' world. This scenario would also help provide support mechanisms for those who were less sure of VRML.

We therefore placed the participants in a 'real' space that they would feel familiar with, an approximation of an art and design studio. A clay workshop was commandeered and cleared and the space broken up with dividing boards and desks forming a variety of re-arrangeable niches for computers and people. We hoped that the inherent protocols of what space was private and what was public in this 'real' space would be transferred to the 'virtual' space.

Layout of the virtual world

To provide continuity we decided that the virtual environment should reflect the real and that we would like participants to be able to enter a global public world and then proceed into their own or another's private space. We wished participants to be able to enter their space by moving into the proximity of some kind of 3-D representation of their individual space, rather like moving around the real studio and then entering the divisions of space that belonged to themselves or others.

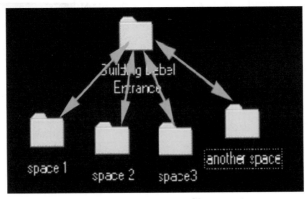

Figure 21: Workspaces file structure

Whilst the participants were actually in one room limited by walls, in virtual space the doorways to other spaces could be anywhere, the only restrictions being those of navigation. At this point the real and the virtual began to diverge. We found that in terms of navigation the simplest arrangement of the 'gateways' was to place the 3-D icons at ground level, in a line immediately in front of the viewer as they entered the space. The gateways themselves were originally intended to be a reduced version of each individual's space, so reflecting the conceptual basis of the whole structure. It became apparent during the workshop that people were creating far larger and complex worlds than we had anticipated so this facility was altered on the second day. A separate file was created in which participants could place a small representation of their space.

When constructing the virtual environment we tried to bear in mind the fundamental point that the development and installation of each participant's idea should be as simple and as quick as possible, bearing in mind their differing skills and experience. Considering this at the most basic level, using the CVE required that each user be able to:

* create their world
* upload their world file
* view the world

Our solution was a layered environment which we tried to keep as simple as possible. On the desktop, participants were provided with a text editor and various VRML builders such as Spazz 3D and Cosmo Worlds in order that they had as wide a choice as possible to create their world. Once the participant had constructed a VRML world they were then faced with uploading their file. They did this by naming their file as the default world and uploading it via a provided CGI script. This file space mimicked the real world as their folders were contained within a global directory named Babel. Once uploaded the participant could view their creation in the Blaxxun client.

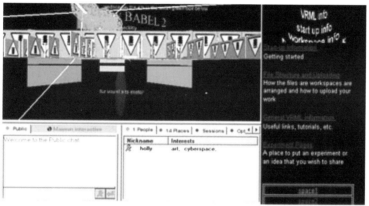

Figure 22: Screenshot of the Building Babel II interface

Although their worlds would be three-dimensional, because of the current state of web technologies, the nature of support materials and the differing skills and familiarity of the people involved, the Building Babel interface was predominantly two-dimensional. Accompanying each 3-D working space and the main Babel environment, in an adjacent frame, we provided a set of files with examples, information about Babel and construction information. The graphical layout of the 2-D files was kept as simple as possible and a black background was chosen so that it would not visually dominate the interface. It was anticipated that towards the end of the workshop, or at some point following the workshop, individual participants would replace the construction information with a file, perhaps describing themselves and their VRML world, and a template was provided for this.

The Event

Day one was a very tiring day and explanation of the technology itself took up much of the time. Participants arrived mid-morning and were given a general introduction to the tools and the interface that they would be working with. We considered that it was very important for the participants to begin building their VRML as quickly as possible and so they were encouraged to make even the simplest piece of geometry and upload it to the space they had chosen.

Choosing a space had no guidelines attached in either the 'real' or the 'virtual' environment. In both more spaces were provided than were actually required, so that the individual did not feel forced into collaboration unless they chose it.

By day two participants began to understand the limits and possibilities of the virtual spaces and software and gained some confidence in the implementation of their ideas. Those with less experience tended to construct less complex ideas and to stay away from complex scripting. They tended to work in an organic manner responding to the tools that they were using rather than implementing a fixed idea.

At the end of day two we gathered to review the progress made. It soon became clear that, whether scripting or not, some of the participants did not have an understanding of the difference between a collaborative and a stand-alone VRML world or, if they did, they did not create a work which took advantage of the unique aspects of a CVE, namely shared events. As the server dealt with the addition and subtraction of avatars into this space it was assumed by many of the group that all events would be shared.

By day three participants were tired and as they began to implement the more complex elements of their work they hit problems. It was really only at this point that participants began to try to add events to their work. We had underestimated the level of technical support which would be needed to deal with this, even for such a small group. Although most individuals were by now comfortable with creating geometry, the complexities of scripting and passing events meant that much of the support groups' time was spent answering what we had previously considered to be relatively simple questions.

To compound this, technical problems began to occur as the server struggled updating work, and as a result of caching problems on their browsers, participants became confused by what they were looking at.

The event ended on a high, however, with participants all gathering to look at the work that had been created. Slowly, shouting around the room stopped and typing in the chat box started as people navigated around. At this point the extent of the problems of inline files became clear as the Babel environment became an unintended conglomerate of all the worlds, a kind of enforced virtual collaboration.

Conclusion

In conclusion it is possible to say that we experienced some success with our workshop and, for the most part, the existing technology was the most disappointing element, although we recognised that we were pushing it to its limits. This is not to say that our planning and implementation of the workshop was impeccable, as some aspects undoubtedly will be changed in the future.

In terms of success we felt that the main premise, that we should integrate the real and the virtual in order to facilitate collaboration, worked very well and it is obvious that bringing people together physically was of major benefit to the process. Participants developed an understanding and awareness of the others present and had the chance to associate a real person with a name, piece of geometry and a communication style.